How to Make Maps

The goal of *How to Make Maps* is to equip readers with the foundational knowledge of concepts they need to conceive, design, and produce maps in a legible, clear, and coherent manner, drawing from both classical and modern theory in cartography.

This book is appropriate for graduate and undergraduate students who are beginning a course of study in geospatial sciences or who wish to begin producing their own maps. While the book assumes no *a priori* knowledge or experience with geospatial software, it may also serve GIS analysts and technicians who wish to explore the principles of cartographic design.

The first part of the book explores the key decisions behind every map, with the aim of providing the reader with a solid foundation in fundamental cartography concepts. Chapters 1 through 3 review foundational mapping concepts and some of the decisions that are a part of every map. This is followed by a discussion of the guiding principles of cartographic design in Chapter 4—how to start thinking about putting a map together in an effective and legible form. Chapter 5 covers map projections, the process of converting the curved earth's surface into a flat representation appropriate for mapping. Chapters 6 and 7 discuss the use of text and color, respectively. Chapter 8 reviews trends in modern cartography to summarize some of the ways the discipline is changing due to new forms of cartographic media that include 3D representations, animated cartography, and mobile cartography. Chapter 9 provides a literature review of the scholarship in cartography. The final component of the book shifts to applied, technical concepts important to cartographic production, covering data quality concepts and the acquisition of geospatial data sources (Chapter 10), and an overview of software applications particularly relevant to modern cartography production: GIS and graphics software (Chapter 11). Chapter 12 concludes the book with examples of real-world cartography projects, discussing the planning, data collection, and design process that lead to the final map products.

This book aspires to introduce readers to the foundational concepts—both theoretical and applied—they need to start the actual work of making maps. The accompanying website offers hands-on exercises to guide readers through the production of a map—from conception through to the final version—as well as PowerPoint slides that accompany the text.

Peter Anthamatten is an Associate Professor and Chair of the Department of Geography & Environmental Sciences at the University of Colorado Denver, USA.

How to Make Maps

An Introduction to Theory and Practice of Cartography

Peter Anthamatten

Routledge
Taylor & Francis Group

LONDON AND NEW YORK

First published 2021
by Routledge
2 Park Square, Milton Park, Abingdon, Oxon OX14 4RN

and by Routledge
52 Vanderbilt Avenue, New York, NY 10017

Routledge is an imprint of the Taylor & Francis Group, an informa business

British Library Cataloguing-in-Publication Data
A catalogue record for this book is available from the British Library

Library of Congress Cataloging-in-Publication Data
Names: Anthamatten, Peter, author.
Title: How to make maps: an introduction to theory and
practice of cartography / Peter Anthamatten.
Description: Abingdon, Oxon; New York, NY: Routledge, 2021. |
Includes bibliographical references and index.
Subjects: LCSH: Cartography. | Geospatial data.
Classification: LCC GA105.3 .A67 2021 (print) |
LCC GA105.3 (ebook) | DDC 526—dc23
LC record available at https://lccn.loc.gov/2020034937
LC ebook record available at https://lccn.loc.gov/2020034938

ISBN: 978-1-138-06779-0 (hbk)
ISBN: 978-1-138-06780-6 (pbk)
ISBN: 978-1-315-15842-6 (ebk)

Typeset in Optima
by codeMantra

Visit the companion website: https://clas.ucdenver.edu/cartography-anthamatten

For my students and colleagues at the University of Colorado Denver, whose public-spirited enthusiasm and optimism I cherish.

Contents

Figures

Acknowledgments

I would like to acknowledge one of my key advisors, Phil Gersmehl, with whom I worked on various cartography projects at the University of Minnesota and then briefly at Hunter College in New York back in the aughts. His logical and practical approach to scholarship, particularly the notion that academics can and should do work that *matters*, in both our research and teaching missions, has served as a fundamental guide to my own work. Some years before I arrived at the University of Minnesota as graduate student, he had published a book called the *Language of Maps* with the National Council for Geographic Education (1991). His perspective on mapping set the foundation for my own approach to the topic, which I suspect is manifest in the text. I often think of lessons I learned from him as I progress through my career.

I must express special gratitude to student collaborators who allowed me to borrow some of their ideas and written work to support parts of this book. Specifically, I thank Cody Peterson for his work on critical cartography, and James Raines III for his work on modern media in cartography. Both Cody and James are exceptional young geospatial scientists who are destined to achieve great things. I hope they will consider teaching cartography someday.

The editors I have worked with on this project have been enormously supportive. They have always been a pleasure to work with, and I would recommend working with Routledge to anyone seeking an academic publisher.

We are fortunate to be members of active and engaging community of professional and academic cartographers who continually advance the field. Countless researchers in cartography whom I have never met have taught me an immense amount through their work.

I am particularly grateful for the support of my family, especially Hilary, Anneliese, and Chloë. I can always count on them, which I always treasure.

1 | Introduction

Maps enable us to view our world in terms of *space*—how we are situated in space, and how the features of the earth's surface fit together. As you start reading this book on the topic of maps and how to make them, take a moment to consider the critical role that maps have played throughout the development of human history and society. It is difficult to imagine the world today without them; maps enabled people to effectively navigate across space and explore the surface of the earth. Maps are powerful: they played a critical role in the conquest and establishment of territory, and then in the governance of that territory. The formal organization of space through maps both caused and prevented conflict throughout history. Maps occasionally influence events in quite overt ways, such as in 2010, when errors in mapping led to troops from Nicaragua to accidentally "invade" neighboring Costa Rica (Brown 2010).

These days, people read maps on a regular basis without much thought. City maps, hiking maps, weather maps, road or subway maps, maps of countries or conflicts, and maps on countless thematic topics have become part of the fabric of daily life. With the rise of mobile phone technology and global positioning systems (GPS), we how have *dynamic* maps that seem to magically show where we are in real time, almost as if it were taken straight out of a Harry Potter novel.

The word "map" appears to be derived from the Latin word *mappa*, which means "sheet" or "cloth," a root word that can be found in other modern English words such as "apron" or "napkin" (Ayto 2011, 337). It is significant that the very origins of the word encapsulate the two-dimensional nature of a map, salient to those who work in cartography. The Oxford English Dictionary (Stevenson 2010) identifies the first use of the word "map" in the English language in the sixteenth century. The word is defined as:

A drawing or other representation of the earth's surface or a part of it made on a flat surface, showing the distribution of physical or geographical features (and often also including socio-economic, political, agricultural, meteorological, etc., information), with each point in the representation corresponding to an actual geographical position according to a fixed scale or projection; a similar representation of the positions of stars in the sky, the surface of a planet, or the like. Also: a plan of the form or layout of something, as a route, a building, etc.

There are some important components to this definition. A photograph or image of the earth's surface is not really a map, because it is not "a drawing or representation" and does not involve symbols. While the term "map" is used interchangeably for a variety of other disciplines and purposes—such as mapping the human genome or the galaxy—cartography has traditionally focused on mapping the *earth's surface*, and so here we rely on that traditional definition. A more succinct but sufficient definition from the *Oxford English Dictionary*'s website is "A diagrammatic representation of an area of land or sea showing physical features, cities, roads, etc." (Oxford English Dictionary 2018).

Maps offer enormous analytical power. A map can be thought of as a "macroscope" that enables the reader to visualize, explore, and analyze patterns that are otherwise invisible to the naked eye. As microscopes enabled significant advances in biology by making visible entities that were otherwise impossible to see—cells, bacteria, viruses, and the like—so have maps enabled advances in geography and other fields by enabling us to view patterns across vast areas of the earth, equally impossible to "see" with the naked eye. Maps have ultimately led to revolutionary advances in our understanding of regional- and global-scale processes.

Because of the diversity of ways in which maps are used in contemporary society, they have become an extremely important and powerful tool associated with a broad range of related technologies that facilitate the collection, management, and analysis of spatial data. Consequently, the art, science, and practice of **cartography**, that is, the design and production of maps, has become a critical skill for many twenty-first-century professionals.

The purpose of this book is to provide you with the background, knowledge, and critical skills for practicing cartography effectively and efficiently. The goal is to equip you with the ideas and concepts you need to conceive, design, and produce maps, drawing from both classical and modern theory in cartography. The principles of cartographic design will be given rigorous treatment, beginning with the traditional principles of design and extending to twenty-first-century map forms (which now include three-dimensional, animated, and user-based designs). The final component of the book will introduce you to some of the nuts and bolts of cartographic production, spanning topics such as the acquisition and preparation of data, refining maps in graphical software, and revising cartographic work.

The first part of this book focuses on the concepts and theories of cartography. The first two chapters review the key concepts in mapping and some of the decisions that are a part of every map. This is followed by a discussion of the guiding principles of cartographic design—in other words, how to start thinking about putting a map together in an effective and legible form. Chapter 5 covers map projections, the process of converting the curved earth's surface into a flat representation appropriate for mapping. Chapters 6 and 7 discuss the use of text and color, respectively. Chapter 8 reviews trends in modern cartography to summarize some of the ways the discipline is changing due to new forms of cartographic media that include 3D representations, animated cartography, and mobile cartography. Finally, Chapter 9 provides a literature review of the scholarship in cartography.

The focus then shifts to the *practice* of cartography in the final section of this book. A strong familiarity with data and the ways the data relate to mapping are critical; Chapter 10 covers data in mapping. Modern work in cartography requires conceptual knowledge and skills in both geographic information systems (GIS) and graphics applications, and so Chapter 11 discusses key concepts and terms in this software. Chapter 12 concludes the book with some examples of cartography projects, discussing the planning, data collection, and design process that lead to the final map products. Your ultimate goal may be to equip yourself with the knowledge and ability to produce your own, original maps, and it is often helpful to start with some guidance. Additional material is provided online to guide you through the process of building a map from scratch, using the common GIS and graphics applications, such as *ArcGIS* and *Adobe Illustrator*.

While I have structured this book in a manner that I believe makes sense for new students of cartography, there are countless ways to approach the study of maps and many of the chapters necessarily draw on concepts discussed elsewhere in the book. Consequently, this book may be approached out of sequence; for example, it might make sense to you or your instructor to review projections before proceeding to other facets of cartographic design and decision-making.

In order to base some of the discussions in the text on real-world examples of cartography, the book includes a map gallery, called "Maps from the Wild," in Appendix 1, a collection of maps taken from the "wilderness" of the Internet and other media sources. These maps come from different authors with different mapping goals, selected to demonstrate the variety in themes, types of maps, and cartographic styles that one encounters on a day-to-day basis in professional occupations and daily life. These examples are discussed throughout the book.

I have also included some resources I hope will be helpful for readers looking to start working hands-on in cartography, or with an eye to develop skills that can translate into a career. Appendix 2 contains an annotated list of cartographic data sources. With the availability of a large variety of open-source applications and data, an autodidact with enough motivation, a computer, and Internet access can produce some spectacular maps. Appendix 3 provides a list of guidelines, drawing from the concepts covered in this book, to assist you in building your own maps or evaluating others. Appendix 4 is a list of professional societies that can help you become involved with the burgeoning community of cartographers.

A brief history of Western cartography

The field of cartography has a rich and dynamic history, involving significant artistic and scholarly achievement that deserves its own study, a story treated elsewhere in a comprehensive manner (e.g., Harley et al. 1987, Bagrow 2010). These are exciting times for the world of mapping; it is difficult to overstate the monumental changes that have taken place over the last several decades in the field. This chapter presents a cursory overview of the key themes and trends to give readers a sense of the context of modern cartographic practice.

The innovation and use of maps may have developed naturally as a function of the way human beings perceive and conceptualize their world. **Spatial thinking** or **spatial cognition**—thinking about the location of objects relation to oneself and other objects, as well as how to navigate through space—may have been critical to survival early in our evolution as a species. It is not difficult to imagine how important this kind of thinking must have been to pre-historic humans, who had to navigate through terrain, locate and remember the locations of food sources, keep track of territory (and disputes over it), and find their way back home. Spatial thinking is certainly fundamental to our species' successful past evolution and remains essential to our daily lives. A report on spatial thinking by the US National Research Council report (National Research Council et al. 2005, 51–52) articulates this idea well:

> Everyday life is impregnated with tasks that, on the surface, are routine and trivial. Viewed as problems requiring solutions, these tasks are far from trivial in nature. All of the tasks involve the concept of space in general and various spaces in particular: global time-space, car trunk space, neighborhood space, supermarket space, room space, machine space, puzzle space, garden space, the space of shoelaces, and the space of dress patterns. Many of the tasks involve graphic representations: the time zone diagram in the telephone directory, the web search engine map, the floor plan of the supermarket, the grocery list, the picture of the completed jigsaw puzzle, and the dress pattern.

Perhaps it is no wonder, then, that some research suggests that children think in terms of mapping—using **allocentric** (from a third perspective) rather than **egocentric** (from the perspective of one's self) representations of space from as young as three years old (Nardini et al. 2006). Humankind has used maps to represent and communicate information about space for at least as long as it has used formal written modes of communication. The oldest identified map may have been an engraved block cave in Spain showing landscape features and hunting sites. The map originates from about 14,000 years ago (Utrilla et al. 2009), 7,000 years before the first writing appeared in China, circa 7,000 BP (Figure 1.1).

FIGURE 1.1
A possible early map, from circa 13,660 BP (Utrilla et al. 2009, 101).

Determining *what a map is*, among pre-historical artifacts, is difficult because it first must be determined that (1) the author intended to represent spatial objects, (2) the objects represented are contemporaneous, and (3) the representations are cartographically appropriate (Smith 1987, 61). Take, for instance, a cave drawing of stick figure men and animals: is this an example of pre-historic art, or is there a spatial meaning inherent in the drawing that makes it a map, or at least map-like? Some of the artifacts that may be examples of pre-historical maps are rock paintings in Spain, drawings of humans in huts in Algeria, what may be Paleolithic engravings of riverside dwellings in Ukraine, as well as other representations of features on the earth and in the sky in other places in Northern Africa, the Middle East, and Europe.

John Nobel Wilford, in his book, *The Mapmakers* (2001), divides the history of Western cartography into four eras of development, which include (1) the idea of mapping and early development of cartography, (2) mapping developments with regard to navigation in the Age of Discovery, leading to the foundations of scientific mapping and modern surveying systems, (3) advances and improvements with the introduction of flight and remote-sensing technologies, such as radar, and finally (4) the development of mapping in the context of space flight and computing, which led to the modern forms of cartography.

Early development of mapping

The origins of Western *cartographic thought* or *cartographic theory* can be found in ancient Greece, starting around 600 BC and culminating with the work of Ptolemy in the Roman Empire, circa 140–160 AD. Ptolemy is often considered to be a founding figure for the discipline of geography and the study of cartography. His key geographic work, called *The Geography*, includes the first systematic coordinate system, used to identify important parts of the Roman Empire (Berggren and Jones 2000). There is good evidence that his contemporaries in China were pursuing similar endeavors in mapping, notably a cartographer named Chang Heng, who also devised a coordinate system, followed by Pei Xui, who provided guidelines for systematic cartography (Wilford 2001, 33–34).

Following Ptolemy's time, cartography went into dormancy for several centuries in the West as Europe entered the Middle Ages, following the decline of the Roman Empire. There are fewer than 600 surviving maps from the millennium between 300 and 1300 AD (Wilford 2001). The maps from that time were not true representations of the earth, but were often fanciful, reflecting the perspectives or wishes of their creators, frequently embellished with mythical creatures and stories tied to Christianity. Like all cartographic work, these maps reflect the culture and society of the time. One of the best-known maps from near the end of this era is the *Hereford Mappa Mundi*, located in England (Figure 1.2).

FIGURE 1.2
An image of the Mappa Mundi ("world map"), located in Hereford Cathedral, UK.

The Renaissance, navigational mapping, and emergence of scientific mapping

Fortunately, the classical works of Ptolemy were preserved by Arab scholars through the Middle Ages and then reintroduced to Europe at the dawn of the Renaissance, when they gained some acclaim and interest (Wilford 2001). As European powers discovered the Americas and began systematically exploring other parts of the world, the need for reliable navigation was sharpened, leading to significant accomplishment in the production of nautical charts and maps. The critical role of mapping in exploration, discovery, and, ultimately, conquest ushered a second major era in the development of cartography.

Near the beginning of the Renaissance, Dutch cartographer Gerardus Mercator produced some of the first world maps and developed a projection that continues to be used in modern mapping. Over the next several centuries, mapping and surveying enabled the exploration, settlement, and conquest of much of the rest of the world. During this time, geographers, such as the explorer and naturalist Alexander von Humboldt, made important contributions to cartography (see Figure 1.3).

Around the sixteenth century, maps exploring patterns in themes or topics, rather than tangible features on the earth's surface, began to appear. One of the best-known early examples of analytical cartography was a map by John Snow, a eighteenth-century doctor who produced maps of cholera deaths in the Soho neighborhood in London to support his idea that cholera was water-borne (Johnson 2006). Figure 1.4 shows a reproduction of his map. The black marks represent cholera deaths that occurred during an outbreak. Snow made the compelling case that the deaths were clustered around the Broad Street water pump, arguing that the pump was the origin of the disease.

The introduction of aerial photography and remote sensing

The advent of manned flight, in combination with the ability to produce photographic imagery, yielded obvious benefits to the study of accurate mapping. A Parisian photographer named Gaspard Tournachon, nicknamed "Nader," produced the first aerial photographs from a hot air balloon in 1858 with a wet-plate camera. The development of airplanes and less cumbersome forms of photography in the early part of the twentieth century paved the way for images of large areas of land that greatly improved both the accuracy and the precision of maps.

Mapping was critical to military operations during the two world wars, which provided an important impetus for the refinement of aerial photography, the development of highly precise coordinate systems, and ultimately improvements to the visual design of maps. Following developments in the eighteenth century on the graphical visualization of data (Tufte 1983), atlases dedicated to thematic topics emerged, such as the *Welt-Seuchen-Atlas* (Rodenwaldt 1952).

Modern academic cartography also matured in the twentieth century. The inception of the field was led by a small number of notable individuals in North America (McMaster and McMaster 2002) and Europe (Harley et al. 1987) in the first half

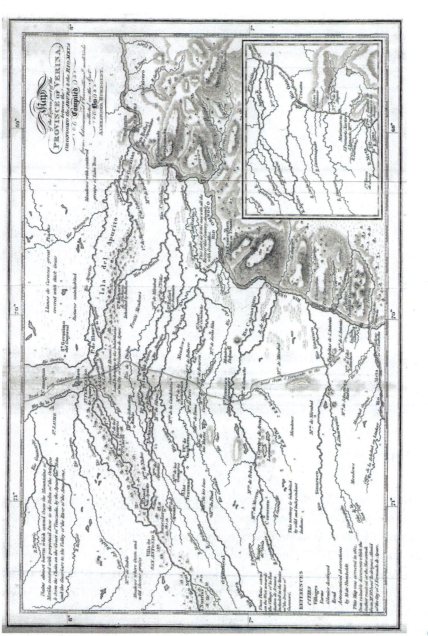

FIGURE 1.3

"Map of the Eastern Part of the Province of Verina, between the Oronooko the Abura & the Rio Meta Compiled from Astronomical Observations & Materials Collected on the Spot by Alexander Humboldt." The image was taken from Princeton University's Digital Collection. (http://libweb5.princeton.edu/visual_materials/maps/websites/thematic-maps/humboldt/humboldt.html).

FIGURE 1.4
John Snow's Map of Cholera.

of the century. During this time, key figures advanced the field by training students who would move to other academic institutions and bring cartography into its maturity. Such figures included John Paul Goode, who produced the well-known *Goode's World Atlas*, currently in its 23rd edition (McNally and Sutton 2016). Erwin Raisz was a Hungarian-American cartographer who produced distinctive maps of landforms and served to organize academic associations around the discipline. Eduard Imhoff, a Swiss cartographer, founded the first academic institute for cartography in 1925.

Individual experiences in the Second World War had an important impact on the field. One of the key participants in the allied mapping effort was Arthur Robinson, who wrote a PhD dissertation titled *The Look of Maps*, in which he argued that mapping should be guided by scientific principles of empirical observation, drawing from research that tested maps and their ability to communicate information (Robinson, Heap, and Roud 1952, MacEachren 1995). Czech geographer Antonín Kolácný similarly pursued the idea of cartography along a communications paradigm, advocating that cartographers should test the results of their work to arrive at optimal mapping solutions. In this way, the study of cartography began to be viewed as an empirical communications science (McMaster and McMaster 2002).

Mapping in the modern era

Several important technologies in the second half of the twentieth century brought mapping to its astounding current form. The development of radar (radio detection and ranging) paved the way for systematic exploration of the Amazon in the 1970s, leading to the development of more sophisticated forms of radiometric remote sensing. The ability to launch satellites into orbit made possible the systematic mapping of the entire planet with extremely high precision. Remote sensing definitively answered questions about the size and shape of the earth. The development of the GPS by the US military, fully operational in 1993 (accurate use of which became available to the general public in 2000), meant that location could be accurately identified in nearly any open space on the planet. Developments in computer technology enabled mapping huge amounts of data and the production of dynamic, interactive, and 3D forms of mapping. Finally, the development of the Internet has resulted in the widespread availability of spatial data and a host of applications for mapping the data.

During most of the second half of the twentieth century, the academic discipline of cartography flourished as individual cartography programs developed with foci on different facets of the field, such as statistical mapping or cognitive research (McMaster and McMaster 2002). The emergence of GIS, however, changed academic cartography as geography programs began to shift their focus to this new computer technology. The previous two decades or so have arguably witnessed more changes to the practice of cartography than in all of its previous history. Traditionally constrained to printed sheets of paper, most maps are now read on computer screens. The rapid growth of interactive maps, animated cartography, and three-dimensional cartography have challenged cartographers to devise novel paradigms and guiding principles for designing maps in these new media.

Cartography's art-science tension and MacEachren's map cube

As academic cartography gained maturity in the twentieth century, several competing theoretical approaches framed research on cartography and the practice of mapmaking. This chapter concludes with a brief discussion about paradigms for thinking about the nature, intent, and purpose of your maps. A more in-depth overview of contemporary research in the field is presented in Chapter 9.

Throughout much of the history of cartography, there has been a basic tension between perspectives of cartography is an *art* or a *science*, which yielded a great deal of discussion among academic cartographers (Krygier 1995). Before there was any organized discussion about the principles of cartography, maps often included artistic embellishment, and the authors were usually more concerned with aesthetic presentation than the succinct and clear conveyance of data (Tufte 1983).

Particularly among academics in North America, the pendulum swung sharply the other way in favor of scientific approaches to mapping following the Second World War. In line with a movement spanning across many fields of scholarship, the *quantitative revolution* gained traction in the field, inspired by technologies that could perform computation with previously unfathomable accuracy and speed. Beginning with Arthur Robinson's call to develop a set of systematic and consistent guidelines, many scholars viewed cartography as a type of engineering problem that should strive to minimize data loss or distortion from the inception of the map idea to map reader's interpretation of the final product (Shannon and Weaver 1949, MacEachren 2004). Even during this time, however, several cartographer-scholars continued to emphasize the artistic value in maps, arguing that maps can elicit emotion with subjective qualities that are difficult to quantify (Keates 1984). Others argued that framing maps as objective representations of reality is naïve, because they are created through an infinite number of subjective decisions (Monmonier 2018) that are embedded within specific social and political structures (Wood and Fels 1992).

Alan MacEachren (1995), a modern cartographer who authored a seminal work in the field called *How Maps Work*, argued for an integration of both artistic and scientific approaches to maps. An artistic approach to the study of maps "is intuitive and holistic, achieving improvements through experience supplemented by *critical examination*" (MacEachren 1995, 9), while scientific approaches are "more inductive and often reductionist, breaking the problem into manageable pieces with the assumption that the total picture… [made] clear by systematically examining each individual part of the process" (*ibid*). I often conduct an informal survey of students after their first year of studying maps by asking them to choose whether they view mapping as an artistic or a scientific endeavor. The split between the two groups is usually nearly equal; as MacEachren and others argue, this may be a false dichotomy. The best approaches to mapping include both perspectives in a complementary and cohesive manner.

MacEachern presents a useful way to think about the guiding principles of map use in the modern context. In his book, *Visualization in Modern Cartography* (MacEachren and Taylor 1994), he describes the distinction between cartographic *visualization* and *communication*, arguing that all maps fall on a continuum that contains facets of both (MacEachren 2004, 356–357). **Visualization** refers to the idea of "making data visible" by translating abstract concepts to a human scale in something that can be viewed (either mentally or physically in a diagram), again evoking the concept of maps as a "macro-scope" or "datascope," which enables us to see patterns that are otherwise buried in too much space or data. Exercises in visualization often have vaguely defined objectives, not focused on any particular application or hypothesis. On the other end of the continuum is the idea that maps can serve to communicate specific ideas or advance a particular argument, drawing from the perspective of maps as a tool to transfer information and knowledge and perspective from the cartographer to the map reader (Figure 1.5).

MacEachren discusses three primary factors that make up a theoretical "map cube." Maps can serve to *present known information* on the one hand, or be conceived as an exploratory project to *reveal unknown information* on the other. Consider, for example, a map designed to explore suicide rates in the UK. It is clearly very difficult to derive spatial patterns by relying on tables of data on suicide rates by county… a public health team might produce a map of rates to examine whether there are any systematic spatial patterns in the country, without any *a priori* knowledge about what they will find. The purpose of their map is not about communicating information but rather about exploring data to build knowledge.

This initial map would not be intended for communication with the public, but for the team of public health workers to "privately" explore their data. The second axis on MacEachren's map cube covers the *public* and *private* realms. Many modern maps are

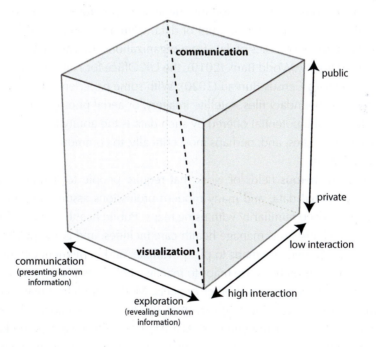

FIGURE 1.5
MacEachren's "map cube" (adapted from MacEachren and Taylor 2013, 6, 1994).

made by individuals or institutions to poke and prod at their own data. They do not intend to use the maps as a tool to communicate spatial data as much as to explore spatial patterns. If the team later wishes to communicate their work or their insights to others, the intent behind their map might shift from the private to the public realms, with implications for how the data are represented.

While the history of cartography is built upon centuries of maps consisting of flat sheets of paper (or other flat media), modern computer technology and geographic visualization software have made it possible to produce *interactive* maps. Users can change what data are shown, how they are displayed or symbolized, standardized, classified, or otherwise manipulated. Interactive map users can now move the extents around to focus on different areas, interactively change the scale of the map, or even change the map projection. Many modern maps continue to be produced on static media, however, such as paper atlases or images on websites. The final axis is about human-map *interaction*, the extent to which the map user can alter the map in real time.

Summary

The ability to read, critically evaluate, and produce graphical summaries of **spatial data**, data tied to locations, is more important now than ever before. We are surrounded by a practically incomprehensible amount of data. With an Internet connection, anyone can access thousands of tables of data from organizations such as the World Health Organization (2020), the World Bank (2019), the UK Office for National Statistics (2011), or the United States Census Bureau (2020). With some experience handling GIS data, users can access boundary files, satellite imagery, or aerial photography. An important key to unlocking the potential offered by such data is the ability to analyze them with geospatial technologies, and, perhaps most critically, to communicate with them effectively through maps.

There are numerous fields of work that require people to handle, analyze, and communicate spatial data, and many modern professions assume that much of their workforce has a deep familiarity with using maps. Public health workers, for example, use maps to track diseases, manage health care facilities, and guide public health initiatives. Urban planners use maps to plan, evaluate, and implement city development projects. Private corporations use maps to manage their logistical operations and identify locations for new stores and other projects. Modern military organizations could barely function without using maps in their day-to-day operations. The Association of Graduate Careers Advisory Services (AGCAS) in the UK identifies the key areas of work for cartographers such as in the context of conservation, government, the military, publishing, and surveying (2016). According to the United States Bureau of Labor

Statistics, the field of cartography and photogrammetry has a "faster than average" growth outlook (2016).

Because cartography is both a discipline of study unto itself and a tool that is used by other disciplines, I encourage you to think about the applications of cartography to the field you are working in or studying. Different areas of work emphasize certain types of maps and often come with their own sets of cartographic conventions that revolve around specific forms of presentation or types of cartographic media. Find examples of mapping in your field or interest, become familiar with the key primary sources of data and cartographic work in that field, and think about how the maps are applied to achieve goals in the field. The rest of this book will focus on the concepts and tools necessary to participate in these endeavors.

Discussion questions

1. This chapter discussed some ways that maps play a role in society and our day-to-day lives. What are other functions of maps can you think of?
2. How do maps affect your life?
3. The text mentions the notions that maps serve as a "macroscope" that enable us to visualize patterns we could not otherwise see. What are some fields of study that benefit from this "macroscopic" perspective and how does mapping contribute to them?
4. Drawing from your own knowledge, what are some ways you believe maps have affected human history and society?
5. There has been a long-standing discussion over the idea that cartography is both an art and a science. Do you believe that it is valuable perspective?

References

[AGCAS], Association of Graduate Careers Advisory Services. 2016. "Job Profile: Cartographer." Accessed August 18, 2018. https://www.prospects.ac.uk/job-profiles/cartographer.

Ayto, J. 2011. *Dictionary of Word Origins*. New York: Arcade.

Bagrow, L. 2010. *History of Cartography*. London: Transaction Publishers.

Berggren, J.L., and A. Jones. 2000. *Ptolemy's Geography: An Annotated Translation of the Theoretical Chapters*. Princeton, NJ: Princeton University Press.

Brown, M. 2010. "Nicaraguan Invasion? Blame Google Maps." Last modified November 8, 2010, accessed July 12, 2018. https://www.wired.com/2010/11/google-maps-error-blamed-for-nicaraguan-invasion/.

Bureau of Labor Statistics. 2016. "Occupational Outlook Handbook: Cartographers and Photogammetrists." Accessed August 18, 2018. https://www.bls.gov/ooh/architecture-and-engineering/cartographers-and-photogrammetrists.htm.

Harley, J.V., J.B. Harley, D. Woodward, G.M. Lewis, M.S. Monmonier, L. de Albuquerque, and J.E. Schwartzberg. 1987. *The History of Cartography: Cartography in Prehistoric, Ancient, and Medieval Europe and the Mediterranean*. Chicago, IL: University of Chicago Press.

Johnson, S. 2006. *The Ghost Map: The Story of London's Most Terrifying Epidemic–and How It Changed Science, Cities, and the Modern World*. New York: Penguin.

Keates, J.S. 1984. "The Cartographic Art." *Cartographica: The International Journal for Geographic Information and Geovisualization* 21 (1):37–43. doi: 10.3138/f27n-7125-8q25-8670.

Krygier, J.B. 1995. "Cartography as an Art and a Science" *Cartographic Journal* 32:3–10. doi:10.1179/caj.1995.32.1.3.

MacEachren, A.M. 2004. *How Maps Work: Representation, Visualization, and Design*. New York: Guilford Publications.

MacEachren, A.M., and D.R.F. Taylor. 1994. *Visualization in Modern Cartography*. Oxford: Elsevier Science & Technology Books.

MacEachren, A.M., and D.R.F. Taylor. 2013. *Visualization in Modern Cartography*. Oxford: Elsevier Science. USA: Elsevier Science Inc.

MacEachren, A.M. 1995. *How Maps Work: Representation, Visualization, and Design*. New York: Guilford Press.

McMaster, R., and S. McMaster. 2002. "A History of Twentieth-Century American Academic Cartography." *Cartography and Geographic Information Science* 29 (3):305–321. doi: 10.1559/152304002782008486.

McNally, R., and C.J. Sutton. 2016. *Goode's World Atlas*. Skokie, IL: Pearson Education.

Monmonier, M. 2018. *How to Lie with Maps, Third Edition*. Chicago, IL: University of Chicago Press.

Nardini, M., N. Burgess, K. Breckenridge, and J. Atkinson. 2006. "Differential Developmental Trajectories for Egocentric, Environmental and Intrinsic Frames of Reference in Spatial Memory." *Cognition* 101 (1):153–172. doi: 10.1016/j.cognition.2005.09.005.

National Research Council, Division on Earth and Life Studies, Board on Earth Sciences and Resources, Committee on Geography, and Committee on the Support for Thinking Spatially. 2005. *Learning to Think Spatially: GIS as a Support System in the K-12 Curriculum*. Washington, DC: National Academies Press.

Oxford English Dictionary. 2018. "Map." Accessed March 3, 2019. https://en.oxforddictionaries.com/definition/map.

Robinson, A.H., H.R. Heap, and K. Roud. 1952. *The Look of Maps: An Examination of Cartographic Design*. Madison: University of Wisconsin Press.

Rodenwaldt, E. 1952. *Welt-Seuchen-Atlas: Weltatlas der Seuchenverbreitung und Seuchenbewegeng*. Hamburg: Falk-Verlag.

Shannon, C.E., and W. Weaver. 1949. *The Mathematical Theory of Communication*. Champaign: University of Illinois Press.

Smith, C.D. 1987. "Cartography in the Prehistoric Period in the Old World: Europe, the Middle East, and North Africa." In *The History of Cartography*, edited by J.B. Harley, David Woodward, and Mark S. Monmonier, 54–101. Chicago, IL: University of Chicago Press.

Stevenson, A. 2010. *Oxford Dictionary of English*. Oxford: Oxford University Press.

Tufte, E.R. 1983. *The Visual Display of Quantitative Information*. Cheshire: Graphics Press.

UK Office for National Statistics. 2011. "2011 Census." Accessed July 7, 2020. https://www.ons.gov.uk/census/2011census.

US Census Bureau. 2020. "United States Census Bureau." Accessed July 7, 2020. https://www.census.gov/.

Utrilla, P., C. Mazo, M.C. Sopena, M. Martinez-Bea, and R. Domingo. 2009. "A Palaeolithic Map from 13,660 calBP: Engraved Stone Blocks from the Late Magdalenian in Abauntz Cave (Navarra, Spain)." *Journal of Human Evolution* 57 (2):99–111. doi: 10.1016/j.jhevol.2009.05.005.

Wilford, J.N. 2001. *The Mapmakers.* New York: Vintage Books.

Wood, D., and J. Fels. 1992. *The Power of Maps.* New York: Guilford Publications.

World Bank. 2019. "World Bank Open Data." Accessed August 22, 2019. https://data.worldbank.org/.

World Health Organization. 2020. "Health Data and Statistics." Accessed July 7, 2020. http://www.who.int/healthinfo/statistics/en/.

2 | **Mapping concepts**

I have often found that casting maps in a language metaphor can be helpful for thinking through and teaching cartography concepts (see Gersmehl 1991). Both language and maps are symbolic, condensed representations of reality. Maps have their own "grammar," a prescribed structure, and a set of rules that creates meaning. If these rules are ignored when designing a map, the final product can end up difficult or impossible to understand. As I discussed in the previous chapter, cartography, like literature, has a distinguished history characterized by significant artistic and scientific accomplishment. Both written text and maps can convey a sense of authority to the reader that can mislead readers, particularly uncritical ones.

A common first step to becoming fluent in any language is to learn some vocabulary—basic symbols that you put together in such a way so as to convey meaning to others. This chapter serves as a "vocabulary builder" as you strive to become fluent in the language of maps, starting with the classes and types of common maps. Following that, I discuss three of the fundamental attributes of a map (Monmonier 1996), namely, the **scale, symbolization**, and **projection**, and then review the data concepts of levels of measurement and classification, important in building thematic maps.

Types of maps

Maps serve a variety of purposes and can assume a huge variety of formats. Having a good understanding of the different types of maps and approaches to representing data is key to the practice of cartography. Throughout the discussion that follows (and in other parts of the book), the text refers to a small collection of "Maps from the Wild" in Appendix 1, examples of maps taken from the "wilderness" of the mapping world, to demonstrate the principles of cartography and cartographic design. The purpose of this map gallery is to prompt you to think critically about real-world examples of maps, each of which was designed with a specific audience and communication goal. I recommend that you take a moment to look over these map before you read the following discussion.

Reference maps

The first thing that most people think of when they hear the word "map" is a **reference map**. Reference maps are not guided by any specific topic or theme but rather provide a general overview of a place or region, usually intended for general interest or navigation. If you are unable to immediately tell what a map is about, it is probably a reference map.

Common examples of reference maps include country maps, road atlases, or the search results on Google Maps. Figure A.4 from the map gallery shows a general map of Europe from 1911. National territories are symbolized through different colors, mountain ranges are represented by shading, and the map includes major cities, rivers, and other water bodies.

Many countries now produce and publish large-scale maps of **topography**, reference maps that show the shape of the land and various other features. A well-known topographic map series is the United States Geological Survey's (USGS) topographic quadrangle maps. The US government has been systematically surveying land to construct these maps since 1884. The maps include **contour lines**, which are lines that show points of equal elevation, to represent the terrain, as well as other features of general interest, such as roads, rivers, building footprints, and political boundaries. Starting 2009, the USGS began transferring its topographic maps to digital format, and you can now download and view the maps from its website (ngmdb.usgs.gov/topoview/viewer/). The UK has produced a comparable series of maps called the UK Ordnance Survey. Figure A.6 shows an example of excerpt of an Ordnance Survey map near the Wash in Lincolnshire.

Thematic maps

Thematic maps are designed with a specific purpose in mind. They are often intended to explore the spatial distribution of a selected phenomenon in the context of a larger project. Figure A.1, for instance, is an example of a map of the world that shows the under-five mortality rate for children by country. The purpose of the map is to examine patterns in health by exploring mortality rates. Some patterns in mortality immediately become clear: rates are low—less than 10 deaths per 1,000 births—in the wealthiest parts of the world, including Europe, North America, Japan, and Australia. Middle-income countries, in places such as Latin America and Asia, experience moderate levels of mortality, while central parts of Africa experience the highest rates. Like any well-designed map, a lot of information not critical to the task at hand is omitted. The map does not include major cities, rivers, or labels for the major bodies of water. It does, however, include everything the reader needs to see patterns in mortality symbolized by country to show the mortality rates. As such, it does a reasonably effective job achieving its purpose.

There are numerous types of thematic maps, often defined around the main type of the symbol used. While grouping thematic maps into subtypes is a useful exercise for thinking of approaches to building them, all these are ultimately artificial designations, as there are countless ways to apply symbology. The following section discusses examples of some common types of thematic maps.

If the purpose of a map is to show how discrete phenomena are distributed across space, a **point distribution** (or "**dot distribution**") **map** might be an effective choice. Point distribution maps merely show how a set or a class of points are distributed. Figure A.2 shows how French toponyms (or "place names") are distributed across the state of Minnesota in the USA. Named places are symbolized by gray points, and the place names with French origins are shown in red. When you view this map, you can immediately see how place names of towns and cities are distributed across Minnesota. The primary goal is to communicate patterns in the distribution of the French toponyms, which appear to be grouped around major rivers and waterways, reflective of the period of history when the area was exploited by French traders, who relied on the water network for their trade. As the state developed over time and settlements moved away from the major waterways, other groups of newcomers (particularly English-speaking and Scandinavian immigrants) chose place names from their own language as French toponyms became less common.

A **graduated symbol map** or **proportional symbol map** also uses points as the chief thematic symbol, but the purpose is to demonstrate the magnitude of some theme or topic by varying the size of the points. Figure A.8 shows reported cases of tuberculosis by country in 2008. The points are centered around countries, for which number of cases are reported. The number of cases at each point are communicated by varying the size of the point—the larger the size of the circle, the higher the number of cases. The map highlights the problem areas where tuberculosis is prevalent in southern Africa and Southeastern Asia.

If you want to highlight the flow or movement of something, such as the flow rates of rivers, the movement of goods in a trade network, or economic migrants to new countries, a **flow map** might be a good choice. A flow map can show how different regions are connected through the movement of something and can also communicate the flow rate by varying the size of the lines.

Isopleth or **isoline maps** use the same technique as contour maps but use isolines to represent an appropriate geographic phenomenon. The areas between the isolines can be shaded to highlight different values, or the lines themselves can be labeled. This type of map is useful for showing continuous data that cover the entire surface of a mapped area. Common uses of isoline maps include topics such as temperature, precipitation, or ecological biome. Figure A.9 shows an isopleth map of Asia, depicting Asia of a late summer night.

Choropleth maps have become an especially common type of thematic map over the last several decades. Choropleth maps use color or shading to show data about statistical-administrative units (countries, states, counties, provinces, etc.). Because the maps are based around *a priori* political entities, the boundaries often have little to do with the thing being mapped. However, they are often quite easy to produce; there is a lot of statistical-administrative data available from governments, and so they are a convenient way to examine a range of geographic patterns. Figure A.11 is a choropleth map of heart disease among women based on data from Health Service Area in the United States. The map has essentially transformed health statistics from each of the administrative areas into a legible summary that shows the key patterns in mortality rates across the country.

One key point to remember when reading or creating choropleth maps is that it is often critical to **standardize** the data appropriately. If you examined *counts* of deaths related to heart disease in the USA, for example, the areas with the largest populations, such as in New York or California, would show up on the map with the highest numbers; those two states have very high populations and so we expect to see more deaths there. However, we do not want the map to highlight patterns in population; we are interested in heart disease. A basic solution is to ensure that the maps show a **rate**, rather than a count, so that the pattern reveals something beyond the size or population of the area. The **crude rate** is the number of cases divided by the total population, yielding a figure such as "number of deaths from heart disease per 100,000 population."

Even the crude rate can be problematic. If you examined a map of cancer mortality rates by state in the USA, you would see extremely high values in some states, such as Florida in the extreme southeast (Figure 2.1).

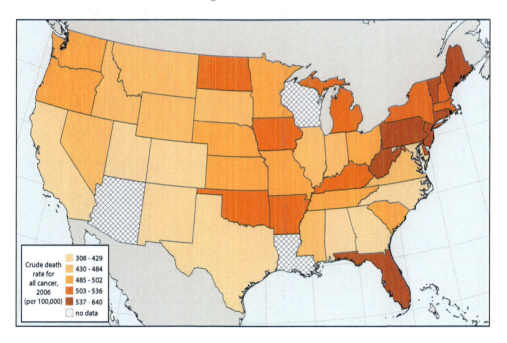

FIGURE 2.1
A simple choropleth map of crude death rates for cancer by state in the USA.

Does this suggest that Florida is a particularly unhealthy state to live in, or is something else going on? The problem is that the age structure of the states varies tremendously. Florida is a popular destination for retirees in the USA and attracts large numbers of elderly migrants. The states along the East Coast similarly have relatively elderly population. This map ultimately reflects patterns in the age structure rather than the phenomenon we would like to explore (cancer rates). We can standardize for the data by performing **age adjustments** on the mapped data, which removes the effects of the population structure by applying the same age structure to each state (Figure 2.2). The age-adjusted version is a better map for the purpose of exploring cancer rates because it removes age from the equation, which is an important driver of that pattern. Figure A.11 also presents age-adjusted data.

A creative way to get around the standardization problem is with a **cartogram**. Cartograms distort the areas and shapes of the mapped areas to show the relative importance of each of the units in terms of some attribute. Figure A.10 contains several maps of US voting patterns in 2016 presidential election. The map on the top left shows how the majority of the population in each state voted in the election—the majority of the population in the red-colored states voted for the Republican Party candidate Donald Trump and the blue-colored states voted for the Democratic Party candidate Hilary Clinton. In the map on the lower left, the area of each of the state is distorted to show

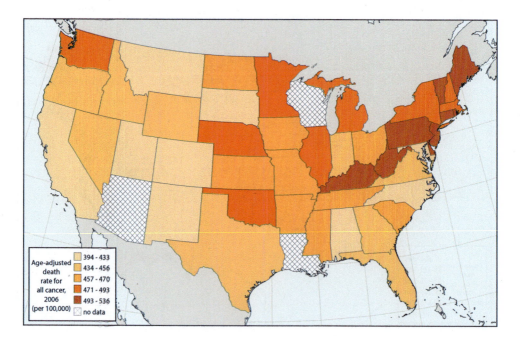

FIGURE 2.2
An age-adjusted choropleth map of crude death rates for cancer by state in the USA.

the size of the total electorate, presenting quite a different story. The maps in the middle show the same for counties. The maps on the right add additional complexity by varying color hue to show the direction of the vote. The color is scaled on a red-blue continuum, with counties expressing a stronger majority for the Republican candidate shown as red and those with a stronger preference for the Democratic candidate shown as blue.

The advantage of cartograms is that by rescaling the sizes of the areas, it compels the reader to think about the importance of each of the counties in terms of their voting power—the critical factor in an election—rather than the physical size of the territory. Cartograms can serve as an excellent tool to make important points and highlight geographic patterns, but it assumes that the audience has quite a bit of familiarity with the region in the map.

Map scale

All maps are ultimately *models* of the phenomena they represent; any distance on a map should represent some distance in the real world. Related to the concept of **scale** are some important terms: **Map distance** is the literal distance on the map, which you can measure in centimeters or inches. The **real-world distance** or **ground distance** is the distance that it represents in the real world, such as kilometers or miles. Most good maps report the scale in some form to communicate that relation to readers.

Translating map scales into real-world distances is usually a straightforward process, although it can be confusing for people not accustomed to working with fractions or large numbers. Suppose that you wanted to estimate the ground distance between two cities on a map with a scale of 1:5,000,000, which means that every unit of map distance represents 5,000,000 (five million) units of ground distance. On the map, the cities are drawn 7.5 centimeters apart. Since you know the map scale, you can calculate how many centimeters this represents in the real world: 7.5 * 5 million = 37.5 million centimeters. A distance measure like "37,500,000 centimeters" does not make a lot of sense to most people, so you need to convert it to a familiar form of geographic distance. We know that there are 100,000 centimeters in a kilometer and can therefore estimate that the cities are 375 kilometers apart (37,500,000 centimeters / 100,000 = 375 kilometers)—at least according to the map.

Cartographers frequently confuse their non-cartographer counterparts with the terms "small-scale" or "large-scale" because they are referring to the representative fraction rather than to the area that appears on the map. When a cartographer uses the phrase "small-scale," they are thinking in terms of **cartographic scale**. Figure A.1 shows the entire world on a single page, with a representative fraction of about 1:250,000,000 (250 million). The fraction (1/250,000,000) is a tiny number. Figure A.2 uses about the same amount of map space, but represents a much smaller area. Its representative fraction is

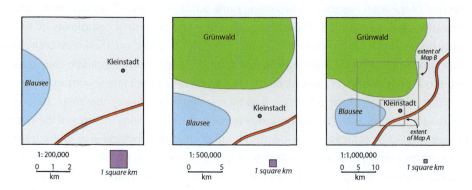

FIGURE 2.3
An example of a map at progressively smaller map scales.

about 1:5,000,000. Because 1/5,000,000 is a larger number than 1/250,000,000, the map of Minnesota is a "larger-scale" map. This usage directly contrasts the way the word "scale" is used in common parlance. When people talk about a "large-scale" operation, they generally refer to something that covers a lot of material or physical space. Because cartographers refer to scale in this way, it is important to be aware of the usage of the term in the field (it is also helpful to be aware of how easily the term can be confused when you are communicating these ideas to others!)

Figure 2.3 shows a basic example of maps of a fictional area at different scales. Map A is the *largest*-scale map that shows the *least* amount of area, the smallest **extent**. Maps B and C show progressively smaller scales, showing more land area using the same amount of map space on the paper. Consequently, Map C has a much larger extent, showing 25 times the amount of area as map A.

Notice that in these maps, with simple symbols and few features, it is difficult to estimate how much area is represented on these maps based on the symbology alone, and so including a scale underneath each map is critical. As you learn about the different map elements and types of cartographic decisions, you will soon discover the choice of scale affects many of the other decisions; you should always consider the implications of the scale of the map for how you symbolize features.

Symbolization

MacEachren starts his seminal book on cartography, *How Maps Work*, with the words "cartography is about *representation*" (MacEachren 1995, 1, emphasis mine). Maps are about representing and communicating real data in a way that is logical and comprehensible to the reader. Consequently, it is important to think carefully about both the *data* and the **symbols** we can use to represent the data. Making effective use of symbols requires both the creator of the symbols and the reader to have a clear understanding of the phenomena they represent. The words on this page are examples of symbols; you have learned how the abstract, squiggly lines that make up these letters translate into sounds, which can be put together into words, which are codes for meaning.

While spoken languages benefits from a well-established and sometimes strictly codified set of rules to govern its symbolization, the language of maps is not so clear-cut. The application of visual symbols in cartography is not random; however, there are some rules that make logical (often "common") sense, as well as some that have resulted from years of use and cartographic convention. To start thinking about symbolizing data in maps, it is important to clarify some of the types of symbols and the different ways they can be used to represent reality or some other idea. Symbols can be defined first and foremost by their **dimensionality**, divided into four basic classes: *points*, *lines*, *areas*, and *volume*.

Points are often used to represent phenomena such as cities and towns or features such as landmarks. In the map gallery, point features are used to represent cities and towns in several of the small- to medium-scale maps (Figures A.2 through A.6). Notice the role of scale here. In Figure A.5, which shows United Nations military deployment

FIGURE 2.4
A screenshot of a map of London using volumetric symbols from Google Earth. Google and the Google logo are registered trademarks of Google Inc., used with permission.

in Lebanon, points are used to represent all the cities, but the cities large enough to be legible as areas (such as Marrakah) are displayed as a light brown color behind the points and labels for the cities.

Lines are often used to show rivers, roads, boundaries, paths, as well as the graticule (the lines of latitude and longitude). Areas can be used to show political territory, forest coverage, classes of geology, or any other feature with an aerial extent. Volume has emerged more recently in maps that are presented in three dimensions (or which mimic three dimensions on a flat screen, such as Google Earth's three-dimensional renditions of city scales, as shown in Figure 2.4). Ultimately all physical phenomena (including points on a map) take up layout space, and so the decision behind which type of symbol to use is often driven by the map scale.

Types of data

The data-driven nature of modern mapping has led many university cartography programs to require students to have a course or two in math and statistics. Because mapping is about collecting, transforming, and ultimately communicating different forms of data—some even argue that maps *are* data—it is important to have a basic understanding of data concepts and terms pertinent to mapping. While you do not need a degree in spatial statistics to produce good maps, mapping requires a lot of data handling and careful treatment of data. Some of the most egregious examples of bad cartography and dishonest map-making are borne from a poor understanding of data and what they entail. The types of symbols and visual variables used to represent features in a map should be chosen after consideration of the map's purpose and the type of data that it aims to represent.

Levels of measurement

Often the first step in thinking about how to represent data is to carefully think about the data type. We can start by thinking about data in terms of its "level of measurement," a concept introduced in 1946 by a Harvard psychologist (Stevens 1946) and commonly applied in statistics. The most basic scale of measurement is nominal, which reports a quality of the data rather than a numerical measure. The word "nominal" means "named," so you can think of data that are distinguished by a name only rather than by quantity. Examples of nominal data include "language spoken" (e.g., English, German, Amharic, or Chinese) or "type of forest" (e.g., tropical, temperate, boreal). An easy way to identify nominal data is to think about whether it is possible to rank it logically—there is no inherently logical way to rank languages or types of forests, for example. Because nominal data are really about the "quality" of data, they are often referred to as qualitative data.

Quantitative data, on the other hand, can be ordered in some logical way. Three levels of measurement apply to quantitative data. Ordinal data can be ranked in logical order, but it is not possible to perform mathematical operations on them. Classifying football teams as the "first," "second," or "third" ranked teams in a league is an example of ordinal data: you can place the teams in a logical order, but you cannot perform any mathematical operations on them. It does not make sense to claim that the third-ranked team minus the first-ranked team is equal to the second-ranked team, for instance. Similarly, geographic concepts such as "village," "town," and "metropolis" can be thought of as an ordinal type of data because there is indeed a clear ranking, but no possibility for performing meaningful mathematical operations.

Mathematical operations can be performed on interval data, which have specific numbers attached to them but lack any "true zero point;" that is to say, "zero" does not mean the absence of the phenomenon. Consequently, it does not make sense to multiply or divide the data. Common examples of interval data include temperature and time. While we can claim that 15 degrees centigrade is 10 degrees lower than 25 degrees, we can hardly claim that a 15-degree day is 60% as warm as a 25-degree day.

Finally, ratio data are often considered the gold standard scale of measurement because they contain a zero value that means the phenomenon is lacking. If we are comparing the populations of two cities, it makes sense to claim that a city with a population of 400,000 has half of the population of a city with 800,000 because the value of "zero" means "no population." Ratio data enable a wide range of mathematical operations and manipulation of the data to make them comparable or easy to understand. While understanding these basic constructs about data can be extremely helpful in a broad range of work in cartographic applications, the key distinction is that qualitative data cannot be logically ranked, whereas quantitative data can.

Discrete and continuous data

Another important quality of data for mapping purposes is the distinction between discrete and continuous spatial data. Discrete data are those which have a clearly defined starting and end point, a presence or absence. A good example of discrete data is water; at any given point in time, every space on the planet either is or is not covered by water. Automobiles, houses, cities, rivers, and people are examples of discrete data. Data for which there is some value at every point in space, by its very definition, on the other hand, are continuous. Common examples include temperature and elevation. There are no points on earth without a temperature or elevation value.

Whether data are discrete or continuous has important implications for how they are symbolized. What symbols should we use to represent continuous data, such as elevation? We could use area symbols by filling in areas with different elevations, using lightness and saturation. Alternatively, we could communicate elevation with isolines or contour lines that connect all the points of equal value. If we have specific elevation data from points around an area, from benchmarks, perhaps, we could also merely include spot elevations with labeled points.

Precision and accuracy

Cartographers often say that "a map is only as good as the data that go into it." The best design techniques cannot yield a useful map if the data quality is poor, and it is often incumbent upon the cartographer to communicate how much confidence they have in the data. We generally prefer to work with data that are **accurate**, true representations of reality. The concept of accuracy is often confused with **precision**, or how specific the data are, but the two concepts are distinct. If you claim that the weather is "warmer than usual for this time of year," your statement might be perfectly accurate, but not very precise. On the other hand, if you claim that "the current temperature is 31.56 degrees centigrade," you are making a much more precise claim (even though your statement may or may not be accurate at all!). **False precision** is a dubious technique that draws from the tendency to assume that high precision implies accuracy. It might be natural for us to assume that if someone reports an extremely precise measure of temperature, they must have an awfully good idea about the actual value.

The concepts of accuracy and precision are extremely relevant to mapped data and cartography. Accuracy in mapping can refer to any of the information presented on the map, including the locations of features, the data about them, or other information presented in the map layout. Precision, on the other hand, refers to the detail of the mapped data; a coastline that shows a lot of fine squiggles that presumably show the true shape and form of the coast is more precise than one that shows the coast as a smooth line. The concept of false precision also applies to cartography; if great detail about data is presented (in the form of highly defined and specific lines, for example, or attribute data reported with a high degree of numerical precision), it implies that the precision is justified by good data. One of the challenges for cartographers is to clearly communicate the **certainty** of the map, or how confident we are about the data, a topic covered in greater depth in Chapter 10.

Data classification

Data classification refers to the methods for placing data into groups. Classifying data is important in cartography, especially for thematic maps. Many maps place data in groups to make the data easy to read and to enable readers to clearly distinguish between categories on the map. How the data are classified can have a dramatic impact on the main ideas of maps, as well as their effectiveness in showing spatial patterns.

For instance, suppose that you were given the task to divide the members of European Union (EU) into income groups and then show the results on a choropleth map. How many categories should you use on your map? When making the decision, you should strike a balance between having enough categories to show some meaningful patterns in the data, but not so many that it is difficult to distinguish between the categories.

In this example, we will divide EU countries into *four* income categories. Table 2.1 shows the gross national income (GNI) per capita figures for EU member states in 2016. The question is then how to divide the countries into four income groups—where do we draw the lines between the categories? There are several key methods that carry assumptions about the data and the reader. You can start by thinking about the purpose of dividing the data. Is the point to show important differences in the groups? Perhaps you think it is more important to divide the countries into quartiles—into equal fourths? Alternatively, you may wish to draw from statistical concepts to show which countries are unusually wealthy or poor, focusing on comparison with the other members of the EU.

The goal for most data classification is to minimize the difference with groups and maximize the distinction between them. The idea is to divide countries into groups that have real meaning; membership in a group implies similarity to the other members of the same group and difference from members of other groups. A traditional, low-technology method for dividing data is to separate the data into *naturally occurring breaks*. The first step is to plot the data to visualize the distribution, perhaps using a histogram (which shows how data fall into ranges or "bins") or point plot (which shows the distribution of the points across a scale). Figure 2.5 shows the point plot for income for the table of EU countries. The numbers on the axis represent the GNI (PPP) of EU member countries from 2017, taken from Table 2.1, in thousands of dollars. Using the **natural breaks** method, you can scan the data to look for points in the data where they seem to break up naturally. In this example, it looks like there is a group of data that bunch up between $25,000 and $40,000, with a break just to the right. There is another clear break around $55,000, and a couple of outliers, two countries with average annual GNI greater than 60,000.

Dividing data into natural breaks is a subjective exercise; different people pick different points to break up the data. This was once a popular and effective method. Good objective measures have since been developed to compute natural breaks. A popular method for mapping applications is called the **Jenks natural breaks algorithm** (Jenks 1967), which serves as the default classification method in *ArcGIS*, the primary commercial geographic information system (GIS) application. With this method, users can specify how many classes they want, and the software will define groups that are as statistically consistent with as much difference between groups as possible.

A second method, **quantiles**, emphasizes the *ranking* instead of the data distribution. To build quantiles, you can start by listing the data in ranked order and then dividing it into equal groups. In the example of income in Europe, there are 28 EU member

TABLE 2.1 This table shows the GNI in international dollars, based on purchasing power parity (PPP) per capita by country for members of the European Union in 2017. The data are in ascending order by value. The Alpha-2 country codes are included as a reference for use with Figure 2.4.

EU Member Country	Alpha-2 Country Code	GNI per Capita (International Dollars)
Bulgaria	BG	20,500
Croatia	HR	24,700
Romania	RO	25,150
Hungary	HU	27,220
Latvia	LV	27,400
Greece	GR	27,820
Poland	PL	28,170
Estonia	EE	31,000
Lithuania	LT	31,030
Slovakia	SK	31,360
Portugal	PT	31,790
Cyprus	CY	33,610
Slovenia	SI	33,910
Czech Republic	CZ	35,010
Malta	MT	36,740
Spain	ES	38,090
Italy	IT	40,030
United Kingdom	UK	43,160
France	FR	43,720
Finland	FI	45,730
Belgium	BE	47,960
Sweden	SE	50,840
Denmark	DK	51,560
Germany	DE	51,760

TABLE 2.1 *Continued*

EU Member Country	Alpha-2 Country Code	GNI per Capita (International Dollars)
The Netherlands	NL	52,640
Austria	AT	52,660
Ireland	IE	62,440
Luxembourg	LU	72,640

Data source: World Bank (2018).

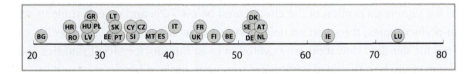

FIGURE 2.5

This is a point distribution plot to aid with visualizing the distribution of the data. The numbers on the axis represent the GNI (PPP) of EU member countries from 2017, taken from Table 2.1, in thousands of dollars.

Data source: World Bank (2018).

countries which we divide into four groups, so we would classify the top seven (since 1/4 of 28 is 7) in the highest-income category, the next seven for the following category, and so on. One advantage of the quantile method is that it is based on the data and it is easy to explain to readers. The map showing per capita income in the EU using this method would show the "top fourth" of countries in terms of income.

The **equal intervals** method forms categories based on the category intervals, usually constructed from the range of the data. In this example, the data show the annual per capita income, and so each interval should span the same range of international dollars per year. You can easily calculate the interval for each category by dividing the range of the entire data set by the number of categories on the map. The country with the highest income in the example data set is Luxembourg, at $72,640, and the lowest is Bulgaria, at $20,500, and so the range is $52,140 (72,640 − 20,500 = 52,140). To make four categories, we divide this data range by four (52,140 / 4 = $13,035), which is the **category range**. We can build each data category by adding that to the previous one. The lowest category would be 20,500 (the lowest income in the dataset) to $33,535 (20,500 + 13,035), the next highest category would start with $33,535 and continue to $46,570, and so forth.

A related method is to define the intervals using "clean" numbers or some other consistent form of categorization. The data in the map in Figure 1A, for example, are broken into classes at 10, 50, 100, and 200, using a consistent progression around numbers that seem to be intuitive or natural breaking points. Schemes that consistently break the data to form "pretty" categories is a common one and can be referred to as "**defined intervals**."

A final method relies on the distribution of the data, or how much the data vary, and is an effective method for clearly representing unusual cases. The **standard deviation** method groups data according to how far it falls from the group average. Standard deviation is a common tool in statistics for describing the dispersion of data, or how much it varies. A set of data with consistent values that are similar has a low standard deviation. As the dispersion of the data points increases, the standard deviation increases. Take an example of a class-wide exam on which the average score is 82.5%. If the scores from an exam all fell between 80 and 85%, the standard deviation would be low. If a significant portion of the class scored below 70% and several students scored near-perfect scores, the standard deviation would be much higher. The specific calculations for standard deviation are covered in most introductory-level statistics courses and can easily be calculated with many computer applications.

This method of data classification uses standard deviations to define the category range. In the example data on income in Europe, the standard deviation is $12,785. Each category can be labeled as "standard deviations above" or "below" the mean. The advantage of this technique is that it highlights the variance of the data and emphasizes the points that are outliers, the ones far above or below what we would expect them to be, given the structure of the data. When cartographers use this method, they should be confident that this kind of classification makes sense for the data and their readers (or otherwise be prepared to describe it in such a way that makes sense to people unfamiliar with the term). Without a basic familiarity with descriptive statistics, the phrase "standard deviations from the mean" probably does not mean much.

The choice of how to categorize data ultimately affects the appearance of the map and what patterns are highlighted. Figure 2.6 is a series of point plots that shows how the national incomes of European countries can be divided into groups with methods discussed above and Figure 2.7 shows the corresponding choropleth maps. While the data are the same, each classification scheme gives visual emphasis to different patterns and groups in Europe's economic landscape.

Summary

Now that you have reviewed some of the fundamental concepts in mapping, such as the types of data that can be mapped, the symbols to represent them, and the key classification schemes, perhaps you can begin to appreciate the vast number of decisions

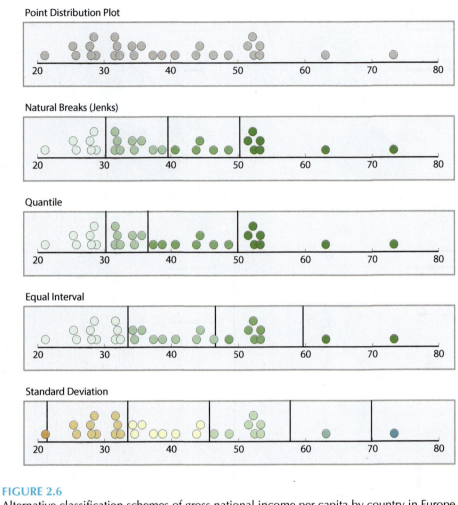

FIGURE 2.6

Alternative classification schemes of gross national income per capita by country in Europe.

Data source: World Bank (2018).

that go into every map. Making a map is similar to writing in some ways; with the ideas, the story, or the information you want to communicate on hand, you must employ the tools at your disposal to translate them into an effective representation. You must thoroughly and thoughtfully edit the information to make it engaging and understandable to your listener or reader.

Some writing tasks (such as a report on your activities at work or a summary report of an article for a college class) come with strict guidelines for production, while other tasks, such as writing a fictional novel, rely on your own imagination and creativity. The guidelines, principles, and tools for cartography function in a similar way. In the next chapter, I will discuss the fundamental decisions you must make when designing a map, and building a toolset for telling your story through mapping.

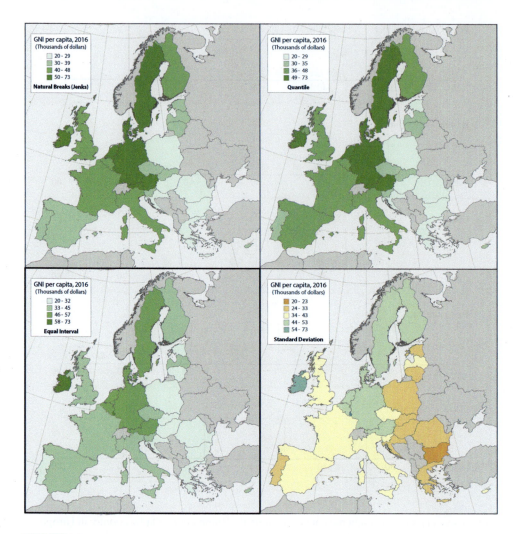

FIGURE 2.7
Choropleth maps of gross national income per capita by country in Europe using alternative classification schemes.

Data source: World Bank (2018).

Discussion questions

1. Can you extend the language analogy to mapping? In what ways can cartography be compared to language?
2. Examine the maps in the "Maps from the Wild" gallery. Can you classify each of these maps into a type of reference or thematic map? What kinds of thematic maps are included?

3. Drawing from the map gallery or other map examples, think about the symbology decisions. Could other symbols have worked effectively for the phenomena being mapped? What are some other features that could improve the maps, and how would you symbolize them?
4. Try to determine what levels of measurement comprise the data on the maps from the map gallery. Are the data discrete or continuous? Can you think of how one might collect or reframe the data to comprise a different level of measurement or type of data?
5. Beyond what is discussed in the text, what are some situations that could call for a particular type of data classification? Is there a type of classification that you find helpful for communicating data? What are some other advantages or drawbacks to these classifications schemes?

References

Gersmehl, P. 1991. *The Language of Maps*. Washington, DC: National Council for Geographic Education.

Jenks, G.F. 1967. "The Data Model Concept in Statistical Mapping." *International Yearbook of Cartography* 7:5.

MacEachren, A.M. 1995. *How Maps Work: Representation, Visualization, and Design*. New York: Guilford Press.

Monmonier, M.S. 1996. *How Lie with Maps*. Chicago, IL: University of Chicago Press.

Stevens, S.S. 1946. "On the Theory of Scales of Measurement." *Science* 103 (2684):677–680. doi: 10.1126/science.103.2684.677.

World Bank. 2018. "GNI per capita, PPP (current international $) – European Union." Accessed July 14, 2018. https://data.worldbank.org/indicator/NY.GNP.PCAP.PP.CD?locations=EU.

3 | The language of maps

Introduction

As one must gain a certain mastery of language to truly use it effectively—by understanding the meanings of specific words and their nuances, how to put those words together in a way that is intelligible to others, and how to use language to make a compelling or insightful argument—so must one master the elements, grammar, and symbolic nuances of maps. The purpose of this chapter is to provide you with an introduction of the basic "grammar" of maps—how to put elements together to tell a cartographic story. In this chapter, I will introduce some additional vocabulary from cartography and will emphasize the kinds of decisions that go into every map. At the heart of this discussion are a few critical and underlying questions that can guide **map design**. Finally, I will review several of the common **map elements** that comprise a map.

The fundamental questions

Initial decisions about what is worth telling should drive how you tell any story. Consider the last time someone asked you a basic, casual question such as: "So, what are you doing at your university?" Your time at a university involves a practically infinite array of details; how you enroll in courses, where you spend your time, what assignments you work on, what you eat for lunch, and what toothpaste you use while there. Some of details clearly do not address the intended purpose of the question and some are not at all important or interesting (at least not to most audiences). Your first step is to determine which sliver of the reality of your studying at the university gets to the heart of the question. An appropriate response might include what your course of study is, how the study is going, or what you have enjoyed doing while there.

You can think about designing a map in the same way you would think about designing a good story—suppose you work for a group or an organization that is interested in promoting good health practices through healthy urban design in your city. You have been given the assignment to produce maps of "walkability," referring to how well the built environment encourages its inhabitants to walk. Examples of good walkability can include well-maintained sidewalks, safe passage across streets, low crime rates, and the existence of plenty of places to walk to. When you are first presented with the challenge of putting a map together from the scratch, the prospect can seem overwhelming. Where do you start? Which features or map elements should you include in a map on walkability? What exactly does your supervisor mean by "walkability," and how can you convey the idea on a map? The question is ultimately how you can make a map that achieves its key objective—to communicate some spatial patterns which inform an audience about the topic.

With so many cartographic decisions to make when you start a mapping project, it is useful to begin by clearly articulating facets of your project to guide your decision-making process. First and foremost, clarify the **map purpose** and what question you hope that it can answer: *What is the purpose of your map?* Write it down or say your answer out loud, if that helps. In this example, you could claim that "the purpose of my project is to communicate patterns of walkability in the city."

This is a good start, but perhaps you should add some nuance to your statement of purpose. The maps you produce are not merely to inform a curious audience about walkability in the city but to guide decision-making and policy by a group of people who may have the power to influence design decisions. Think back to MacEachren's "Cartography Cube" model that articulates the purposes of maps, discussed in Chapter 1. Is the main goal of the map project *exploratory*, as a means of visualizing and highlighting patterns, or is it intended to *communicate* a more focused and specific problem? In this example, the audience is brainstorming about urban planning projects, and so you could think about this as an exploratory exercise. You might therefore amend your map purpose a little to something like "the purpose of my project is to explore patterns of walkability in the city to inform urban design decisions."

Once you have a clear idea about the goal of your maps, the next step is to think a bit about *communication*. Communication is, of course, a two-way process and fails if the recipients of a message do not understand something, which could mean including explanatory text or basic map features. On the other hand, you wish to avoid spending precious layout space with features or map elements that are not necessary. If you know that your audience is extremely familiar with the city, for instance, you can safely assume that they will not need a lot of orienting information, which might give you license to spend more map space on the truly important parts of the main message. Maps of walkability designed for a broader audience—perhaps including people who may never have set foot in the city—would require an entirely different set of components and considerations. A second critical fundamental question is, therefore, *who is the audience*?

After discussing the work with your supervisor, you discover that the intended audience is a group of urban design and public health officials who are deeply involved with planning the urban environment of the city. Included among the committee members are epidemiologists with graduate degrees in public health, seasoned urban planners who are intimately involved with the city and its growth, as well as generally well-educated members of nongovernmental groups and the city council.

At this point, with a clear goal and a basic idea of who you are communicating to, the final question you should clear up with your supervisor is **what is the cartographic medium?** How will these maps be communicated? Will the committee view them in an ephemeral series of slides in a presentation? If the goal is to produce a slide presentation, will these be accompanied by a printed report the audience will be expected to study and scrutinize? Alternatively, will the maps end up in a digital format that the audience can view on the Internet or download? These considerations should drive many of the decisions you will eventually make about the map.

If your audience only has a minute or two during a project presentation to digest the information and the main ideas, then your map will be most effective if your symbols are simple and impactful. You should probably avoid including copious details and focus instead on the main ideas and patterns. You should avoid including much reading—just the main "headlines" of the map, with one or two sentences of text—you want the main ideas to be easily apparent. The physical medium of the map, where it ultimately appears, is also extremely important to consider. Printed maps accommodate different types of colors and symbols from digital maps. Digital maps, on the other hand, could appear on large computer screens, on cell phones, or be projected onto a huge screen in a lecture hall, and so you should plan the design with some flexibility around the final size of the map presentation.

While there are certainly other considerations you can make when you embark on a map-making project (such as how much time you or your team want to spend on the project), these three fundamental components of planning your map are paramount. Clarifying these questions, in combination with some experience and common sense, can serve well to guide the entire process. These fundamental principles can guide the innumerable specific decisions you will make during the mapping process, such as what colors to use, how big to make the font, or how to symbolize the key phenomena in the map.

In summary form, the principal guiding questions are:

1. What is the purpose of the map? What do I hope to achieve with it?

2. Who is the main audience? Who can I anticipate will be the map readers?

3. What is the cartographic medium of the map? Where and in what form could it be viewed?

Map elements

A good story has a cohesive and understandable structure. It presents information in a sensible and clear order that forms a basic framework to guide your audience through the logic of the story. The structure of a map is mediated through its **layout**, a collection of both graphical and written components. The **map elements** are the individual components that comprise the layout; one might think of these elements as the "paragraphs" that come together to make up your story. The best designed layouts use these elements judiciously to make the information easy to grasp, but they can serve a variety of other purposes as well, such as advertising, promoting an organization, attracting viewers to the map, or producing an aesthetically appealing design. In the next chapter, I will discuss some of the principles of good map and layout design. First, however, it is important to build a working vocabulary by learning about the elements in your cartographic toolbox.

As you read and think about these map elements, take a moment to look at the map gallery in Appendix 1 to consider how the map elements serve to orient and guide the reader (you, in this case) to give meaning to the information presented. Try to give these some critical thought—could the map elements be improved to make the map easier to understand and interpret? Are elements missing that you would like to see, or do any of the elements merely clutter the map without adding much information? As you begin to produce your own maps, you will have to consider which elements to include, and each decision will have an impact on the map's legibility, clarity, and overall success. As you will discover when you contemplate your own and others' maps, while there may be strong guidelines and recommendations, there are few steadfast rules in cartography, and nearly all the guidelines have exceptions.

The map

It may initially seem ridiculous to include the map itself among the list of map elements, but it ultimately serves the same role as the other elements of communicating information. Often—though not in every case—the map should take center stage, with the other map elements supporting the map by helping the reader to interpret and understand it. In some cases, the map might be embedded in a larger project or context, serving the mission or goal of that project in some ancillary way. Figure A.12 shows a world map depicting changes in areas protected for conservation across a ten-year period, embedded in a journal article. The map supports the text but is not the key feature of the article.

Title and subtitle

The map title is usually one of the first things that the readers will see, giving them near-instant orientation to the map's purpose. The title should be the readers' first guidance, particularly when the context does not provide an immediate and clear idea about the purpose of the map. When the purpose and content of the map are very clear from the context, the title is less critical. If the map is produced from the results of a search on *Google Maps* (an Internet-based map service), the user is actively interacting with the map and should know what it is about, for instance, and so a title is far less critical. Figure A.9 shows an example of a search result of a weather map on a summer day. The reader, perhaps having searched for or selected a link called "temperature in Asia," should be well oriented to the map's purpose.

A map that is presented as part of a progression of a research report, such as in Figure A.12, may similarly not need a clear or obvious title. When I designed this map, I knew that the readers would encounter it as they were reading our analysis of ecological biases in protected areas, and hence should already know precisely what the map is about.

Map titles can serve to take care of some the map's business, sparing the need for other map elements. The World Health Organization map in Figure A.1 has the rather pragmatic title: *Under-5 mortality rate (probability of dying by age 5) per 1,000 live births, 2015*. This tells the reader about not only the main idea of the map but also what data are included, and it even explains what the "under-5 mortality rate" means. There is little need for much explanation of the legend, since it is already clearly explained by the title.

The legend (key)

A **legend** or **key** of some sort is nearly always critical for an effective map. Most of the maps in the gallery that could be considered "thematic maps" have a legend of some kind. Notice, however, every example here also contains symbols that are not explained. In the first example, *Under-5 mortality rate (probability of dying by age 5) per 1,000 live births, 2015*, there is nothing to indicate what the squiggly lines around the world mean. You probably correctly assumed that they show national boundaries. *French vs. All Place-Names in Minnesota: Simple Point Distributions* (Figure A.2) has unexplained areas and lines of light blue throughout the map which indicate rivers and lakes. Only the university and footpath symbols in Figure A.7, *A Map of University Buildings at Oxford,* are explained in a legend because few readers would have trouble interpreting and using the map without additional guidance. Trees, green space, roads, and buildings are also symbolized on the map without explanation. The authors appear to have had little doubt that the readers would make a correct assumption about what those symbols show; they decided that the symbols would be clear enough for their intended audience and so they do not need to use space to explain their meaning on the map layout.

Figure A.3, *Direct Normal Solar Resource of Missouri,* effectively uses its legend to convey a lot of information succinctly. It shows not only the range of solar resources in Missouri, presented on the map, but also puts the data in context with the rest of the United States, using a legend on the inset. Not only can the map reader see that solar energy resources in the state are concentrated in the Southwest, but also that, in terms of the solar resources for the United States as a whole, Missouri falls quite low on the spectrum.

Cartographic decisions about how to build the legends should be guided by the fundamental questions discussed in the beginning of the chapter. If the cartographer can assume that her or his audience will have seen plenty of political maps of the world, for instance, then little explanation of country boundaries is needed. If including an element serves to prevent confusion or save the reader time interpreting the map, it is usually worth including.

Scale

Including a **map scale** is important for most maps. The scale becomes an even more critical element if the map is intended for navigation, for which calculating distances to travel could be a key part of the map's use. Many people who work in cartography consider scale bars to be non-negotiable, but you might notice that some of the maps in the map gallery lack a scale bar. Is this an important omission from that map? As you consider the question, remember to contemplate the purpose of the map, its audience, and the medium. Would inclusion of a scale bar serve to help the map achieve its goal or clarify important information for readers?

Map scales can assume different forms. The most common form, and perhaps the most intuitive to read, is a **graphical scale bar**, which shows units of distance on the map in terms of the real-world distance (Figure 3.1).

A verbal scale uses text to report the distance, such as "one centimeter on the map represents one kilometer" or "one inch to four miles." This is a basic and intuitive way

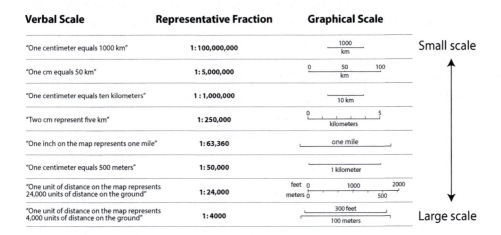

Verbal Scale	Representative Fraction	Graphical Scale	
"One centimeter equals 1000 km"	1:100,000,000	1000 km	Small scale
"One cm equals 50 km"	1:5,000,000	0 50 100 km	
"One centimeter equals ten kilometers"	1:1,000,000	10 km	
"Two cm represent five km"	1:250,000	0 5 kilometers	
"One inch on the map represents one mile"	1:63,360	one mile	
"One centimeter equals 500 meters"	1:50,000	1 kilometer	
"One unit of distance on the map represents 24,000 units of distance on the ground"	1:24,000	feet 0 1000 2000 / meters 0 500	
"One unit of distance on the map represents 4,000 units of distance on the ground"	1:4000	300 feet / 100 meters	Large scale

FIGURE 3.1
Examples of different types of map scales.

to explain map scale; people are generally familiar with terms such as "centimeters" or "inches" for personal space, as well as geographic units of distance such "kilometers" or "miles." However, people have a poor sense of what those units of distance actually mean and are not very good at translating them... you can experiment a bit by asking someone to estimate the distance between a couple of features on a map that uses a verbal scale. Nonetheless, a verbal scale can be useful to convey a general sense of the scale using accessible terminology.

Perhaps the most difficult type of scale for novice map readers to understand, but commonly used in cartography, is a **representative fraction (RF)**, which shows the relation between map space and real-world space in the form of a ratio. Take an example of a map in which one centimeter represents one kilometer; we can think about how the two are proportionally related. Since there are 1,000,000 (one million) centimeters in a kilometer, the representative fraction would therefore be 1:1,000,000, meaning that one unit of distance on a map represents one million of that unit in the real world. Key government map series, such as the maps of the Ordnance Survey produced by the Royal Geographic Society of the UK (Figure A.6 is an excerpt of such a map), are commonly identified by their representative fraction (e.g., the "1:25,000 Ordnance Map Series"). It is important to bear in mind that representative fractions are unitless and that any unit of distance can be applied to them. In a 1:25,000 scale map, one centimeter represents 25,000 centimeters, as one foot represents 25,000 feet.

With the rise of digital maps in recent decades, verbal scales and representative fractions have become less common and useful. Reporting the scale of a map as "1:25,000" can lead to poor results if anyone changes the size of the map—perhaps the digital map you designed on your computer screen is projected on a large wall for presentation or viewed on a mobile phone, for example. The first part of the ratio, the map distance, has changed, but, of course, the real-world distance has not. Graphical scale bars are usually the best choice for modern maps, particularly if map dimensions may change when the map is converted into a digital format or otherwise transformed, because the scale changes along with the map itself.

Reporting the scale on a map can be tricky business in the first place. As is discussed later in the book, cartographic projections distort the mapped representation

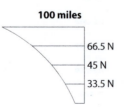

FIGURE 3.2
An example of a variable map scale.

of the surface of the earth. Particularly on small-scale maps that show large areas, the scale *varies* as a function of where it is on the world's surface. In these cases, you might consider using a **variable scale** that shows how the scale changes across different lines of latitude on the map (Figure 3.2).

North arrow

The **north arrow** or **compass rose** should orient the reader by indicating the direction. By convention, north is nearly always oriented at the top of the layout (some geographic software applications do not enable users to easily deviate from this convention). This map element is often considered essential for all maps—I recall learning in primary school that every map should have a north arrow—and people often object if the north arrow is missing.

As you view each map element critically, however, you may notice cases in which it is probably acceptable to leave out a north arrow or compass rose. Notice how many maps from the gallery (Appendix A) do not include them. Most of the world orients north at the top of the layout, and the cartographers have not included the north arrow. The map of toponyms in Minnesota (Figure A.2), in contrast, does include a traditional compass rose. As you evaluate these maps, you can think about whether the lack of a north arrow rose hinders the map's purpose. You probably did not wonder how these maps were oriented.

Another problem with north arrows is that directions are often not consistent across the layout due to the map's projection. The map of Europe (Figure A.4) and the map of tuberculosis around the globe (Figure A.8) show the meridians, the north-south lines of longitude, along which the north-south axis follows. Due to the projection of the map, the direction of "north" on the map varies depending where you are looking on the map, and so a simple compass rose would ultimately be misleading.

There are cases in which a north arrow or compass rose is critical, however. If the cartographer uses a non-traditional orientation—putting south at the top of the map, for instance—it is important that you communicate this to the reader. If determining direction is key to the map, particularly for large-scale maps that may be used for navigation, it is important to include an accurate and explicit guide to the map's orientation.

On maps used for navigation, the north arrow often includes multiple forms of north that might be helpful for users who need precise measures of different forms of "north." Figure 3.3 is an adapted version of a north arrow that appears on USGS 1:24,000 Topographic Quadrangle Maps. "GN" refers to **grid north**, oriented toward the North Pole and showing the orientation of the lines of longitude. The star shows the orientation of *Polaris*, the North Star, and MN shows **magnetic north**. The difference

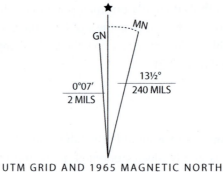

UTM GRID AND 1965 MAGNETIC NORTH
DECLINATION AT CENTER OF SHEET

FIGURE 3.3
An example of a north arrow, as it appears on USGS Topographic Quadrangle Maps.

between the angles is reported both in angular degrees and **milliradians** (equivalent to about .0573 degrees).

Date

I would argue that nearly every map should include a date on it in some form. The world is an ever-changing place, dynamic in ways that are often difficult to imagine. Including the date can convey important information about the relevance of the map's content. Have a look at Figure A.4, for instance. The map, titled, "Europe at the Present Time," does not include a date. Europe clearly looks very different today from how it appears on the map (in fact, the borders changed drastically within a decade of the map's publication, due to the First World War). While the map originally appeared as a page in an atlas where readers may easily determine the date from the context, simply changing the title to "Europe in 1911" would have made the map immediately clearer in a broader variety of circumstances.

Dates are extremely important in any map that represents transient or dynamic data. Population, health, political boundaries, hazards, the built environment, biological features such as forests, and even climate can change dramatically over the course of decades or years, and so including the date becomes a matter of accurate reporting. Maps of dynamic features prone to constant change should include the date prominently in the layout. It is often wise to include both the *date of the data you are mapping*, when appropriate, as well as the *date the map was created*.

While it may seem extraneous to include these dates on maps of relatively static or unchanging features, such as in small-scale maps of geology or topography, for instance, the date conveys information beyond the features being mapped; it can provide the reader with insight about the technology, the state of knowledge about the map's topic, and the context around the general approach and ideas at the time the map was produced.

Authorship

Including **authorship**—the name of the people or organization responsible for the map—shows the reader where the map came from. Having this information demonstrates to the reader that someone is putting their name on the product, which can itself serve to lend the map some credence. Furthermore, map readers can evaluate the credibility of the author themselves. If the organization is a well-known steward of public interest, such as the World Health Organization (such as in Figure A.1), the map gains some credibility by drawing upon the work and reputation of the WHO. Depending on the purpose and intent of the map, I would argue that the authorship generally should not occupy a prominent position on the map but should be legible to most readers. Including the author's contact information can also enable readers to contact you if they have questions about your work, and it gives them a means to seek permission to use your work.

Data sources

Maps are often compiled from a variety of **data sources** that include spatial data (such as boundary files or satellite imagery), as well as tabular attribute data (such as suicide rates by county or country statistics on mortality rates). Including the sources of data on the maps is an occasionally overlooked practice that appropriately attributes others' work, enables readers to evaluate the quality of the data being mapped, and gives readers the option to find the original data for their own projects.

The graticule

The **graticule** is the printed lines of longitude and latitude that comprise the key parallels and meridians in a geographic coordinate system. Including the graticule on a map provides a good general reference about the map's location. In small-scale maps, it also has the benefit of giving savvy map-readers a sense of how the map projection has distorted the surface of the earth. In large-scale maps of smaller areas, including a graticule can be useful for navigation by enabling map readers to pinpoint their locations using other maps or global positioning systems (GPS), such as those common in mobile phones.

Neat lines and frame lines

Some map elements serve primarily as graphical tools to structure the layout and help readers make sense out of it. **Neat lines** are lines around the mapped area that separate the part of the layout that represents some portion of a mapped surface from other parts, such as a textual description or other non-map elements. **Frame lines** are lines around the entire layout. Whether you use these layout elements is often a question of preference. Neat lines and frame lines can compartmentalize the map layout and give it an explicit structure, making the separation of layout elements unambiguous. In my own cartographic work, I tend to gravitate toward compartmentalizing the layout for printed cartography, but I have found that many of my peers and clients prefer more seamless graphical structures. Online Internet maps and maps for mobile phones were once quite heavily compartmentalized but have generally assumed more seamless designs in recent years (Muehlenaus 2014).

Insets

Insets serve two main purposes: to clarify details and to provide context. *Larger-scale* insets highlight detail in a portion of the map where there is so much going on that it becomes unreadable. In a road atlas, for example, there is often plenty of map space to show the main roads in sparsely populated areas, but the road network becomes too congested in towns or cities at the scale of the main map. One solution is to provide an additional, smaller map in the layout that "zooms in" to the congested area, enabling important details to be distinguished.

Smaller-scale insets show the area of the main map with a smaller map that provides some context. In the map gallery, the map of toponyms in Minnesota (Figures A.2), solar power potential in Missouri (A.3), and reference map of Europe (Figure A.4) include examples of smaller-scale insets. The toponym map contains an inset that shows where Minnesota is located within the United States. This inset immediately makes the map more comprehensible for readers who lack familiarity of the states in the USA, but who have a general understanding of the USA and its geography. The solar potential map of the state of Missouri achieves the same thing but also puts the *data* in context. The reference map of Europe (Figure A.4) was clearly designed for a North American audience familiar with the geography of the USA. The inclusion of a small map of Illinois serves to give the readers as sense of the scale of Europe (and some American readers might have been astounded to see that the moderately sized state dwarfs some of the smaller countries in Europe).

Other map elements

Good map layouts can be the product of creative endeavors that take the context and purpose of the layout into careful consideration. There are an infinite number of possibilities and configurations for producing a good map layout. Gretchen Peterson (2014), in her book *GIS Cartography*, provides an excellent overview and practical guide for many map elements beyond the key ones discussed here. Some of these include data graphics, photographs, tables, copyright, and logos, among others. As you consider which among the many possible layout elements to include on your map work, you can think back to the guiding principles: does including a map element help the map achieve its mission and overall objectives, does it assist the audience to read the map, and does it make sense for the cartographic medium (or media) in which the map may appear?

Symbolization and the visual variables

As discussed in the previous chapter, the basic cartographic symbols include points, lines, areas, and volume. This decision about what type of symbol to use might seem straightforward or obvious in many cases, but there are often multiple ways to symbolize the same data, particularly when you work with thematic data. Much of the time, the choice of symbol is driven by the scale and the nature of the phenomena being represented.

Qualitative visual variables

What if the map should communicate something *about* the phenomena it represents? Figure A.2, the map of French and other toponyms in Minnesota, for example, uses points to represent place names, but it also communicates information *about* those names—which ones are derived from French and which ones are not. The map must vary the symbology in some way to communicate that distinction between the origins of place names.

The term **Visual variables** refers to variations in the symbology to communicate information. Good cartography uses visual variables to communicate data efficiently, effectively, and in an aesthetically appealing manner. There is a well-established typology of visual variables and some basic conventions that cartographers should generally follow when using them.

The concept of visual variables was originally introduced by French cartographer Jaques Bertin in his book, *Sémiologie Graphique* (*The Semiology of Graphics*) (Bertin and Barbut 1967). Data about phenomena can be represented by varying the position, size, shape, value (the relative lightness or darkness of a color), color hue, orientation, or texture (the fill pattern) of the symbols on the map. While Bertin's work is not exhaustive and other cartographers have proposed other visual variables, these ideas continue to have currency in modern cartography.

Bertin also provided guidance for how to apply the visual variables. The basic principle behind his work is that symbols that are inherently "rankable," that is, which have a clear ordering, should be used to represent data that also have an unambiguous

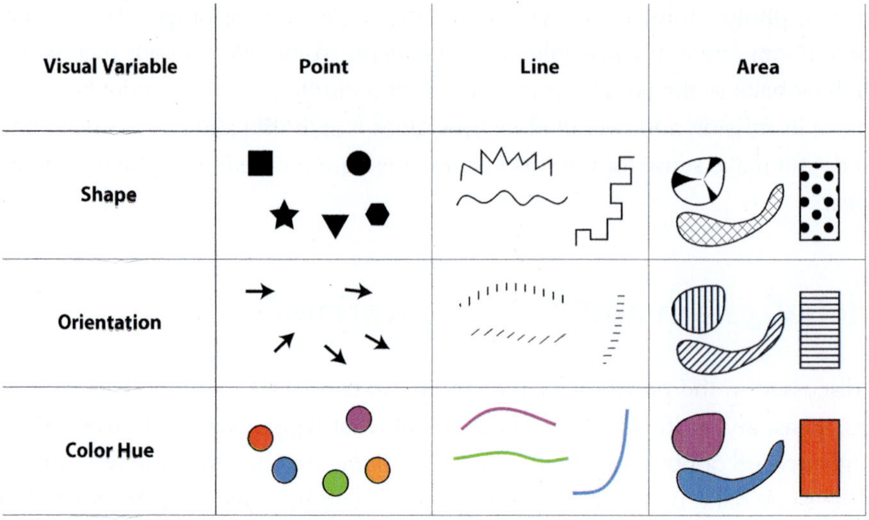

FIGURE 3.4
Bertin's qualitative visual variables: shape, orientation, and color hue.

order. For example, a set of symbols of different sizes—from small to large—could be used to represent the under-five mortality rate by country. Size produces a clear and unambiguous ranking. Similarly, visual variables without any inherent graphical order should represent categorical variables.

Bertin's *qualitative* visual variables include **shape**, **orientation**, **texture**, and **color hue** (Figure 3.4). These visual variables are appropriate for nominal data for several reasons. For nominal data, you should avoid steering your reader into ranking them in any way. If you made a map of "primary language" by city and used the size of points to represent them, the reader is likely to perceive an order that you do not wish to communicate. Furthermore, Bertin argues that these variables are best suited to encourage "selective perception" (2011, 67). The idea is that the reader should be able to easily isolate all elements in a category (to identify all the blue areas on a map, for instance, and immediately recognize that those symbols are in a different category than the red areas).

Clearly, some of these are not effective in mapping. Readers may have difficulty distinguishing between different orientations on area symbols, and other quantitative visual variables, such as color hue, are much more effective. I would generally advise against using shape or orientation with line or area symbols. As you continue to study and learn from maps, you should try to notice how good maps successfully apply visual variables to communicate data.

Consider how these visual variables are applied to the maps in the *Maps in the Wild* gallery. In the early twentieth-century map of Europe (Figure A.4), the author effectively uses color hue to distinguish between different countries ("country" is a nominal type of data). The map of election results in the USA (Figure A.10) also uses color hue effectively to communicate qualitative distinctions in the data; parts of the USA that voted for the Democrat are colored blue (using the traditional colors for the parties in the United States), which are easy to distinguish and compare to the counties that voted for the Republican, colored red.

Quantitative visual variables

The quantitative visual variables include **size**, **texture**, **color value**, and **color saturation**. As you should be able to see in Figure 3.5, there is a clear, inherent order to these; if you presented these symbols to your peers, they would probably rank them consistently. Once again, there is little practical use for some of these combinations—using texture on points or lines is seldom effective, for instance, but the others have been used extensively. With modern printing technologies, it is not often necessary to resort to using texture or size for area symbols, but when printing costs of using colors is expensive (such as was the case for newspapers), size or texture was often an acceptable and reasonably effective option.

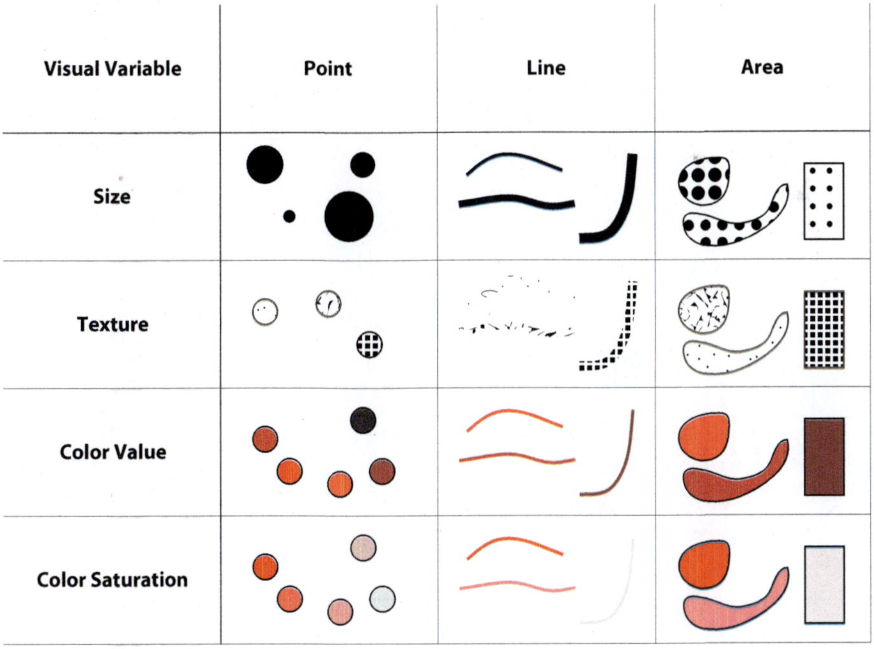

Visual Variable	Point	Line	Area
Size			
Texture			
Color Value			
Color Saturation			

FIGURE 3.5
Bertin's quantitative visual variables: size, texture, color value, and color saturation.

Cartographic conventions and breaking the rules

There are some conventions about which visual variables to use: larger size and brighter, more saturated colors are normally used for higher values. Maps that ignore these conventions can disorient or confuse readers, because most readers have come to expect that order.

A common "error" I often see in mapping is the use of color hue to represent quantitative variables. With a few exceptions, it does not make sense to argue that a blue symbol should be ranked above purple or below green one, for example. Color hue is therefore most appropriately used for data that are qualitative, which differ in terms of their category, name, or quality only, rather than data which are distinguished by a count or quantity. Novice cartographers seem to be seduced by color and are tempted to produce brightly colored maps, but overusing colors often yields maps that are difficult to read.

Consider an example of an inappropriately applied visual variable. Figure 3.6 is a map of the percentage of population with access to the Internet, ratio data with a clear order. While the legend is correctly aligned to the symbols on the map and you can probably distinguish in which parts of the world there is a high rate of Internet access, these decisions force the reader to do a lot of work to decipher the patterns. The reader must learn the color scheme (that red represents the lowest rate, and purple the highest, for example), by taking time to study the legend. Compare this map to Figure 4.2, in

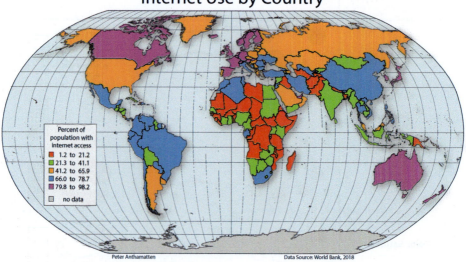

FIGURE 3.6
An example of a choropleth map with a poor selection of visual variables (color hue to represent quantitative data).

the next chapter, that uses lightness and saturation on the same data and layout. The patterns are immediately much clearer, and you do not have to spend more than a few seconds studying the legend.

As is the case with many guidelines in cartography, there are exceptions to the rule. If you have been examining the map gallery carefully, you might have noticed one or two maps that seem to use color hue to show quantitative data. Figure A.9 shows temperature in Asia in the early morning in late summer. The map uses blue color to show cool temperatures, lower than 0°C, yellow to show temperatures between 0 and 20°C, and red to show high temperatures above 20°C. Does this make it a bad map? Do you have trouble reading it? Since using color hue in this way is a common cultural convention in weather maps that people are generally accustomed to seeing, it works well. When you start to make decisions about visual variables in your own cartography, you can always check with your target audience to ensure that the symbol choices are intuitive to them.

Generalization

The key idea Monmonier (2018) makes in his book *How to Lie with Maps* is that maps are vastly simplified models of reality. Not only do maps necessarily offer an extremely selective view that includes but a tiny slice of reality, they must also perform significant distortions in the representation of that reality to produce a usable tool for visualization and communication.

Let us take the example of the map presented in Figure 3.7. This is a modified map of survey results from a research project I worked on around metropolitan Denver, Colorado. This map shows the backdrop (I have not included the survey results in the map). The main purpose of the features is to orient the audience to the area and give them some visual references to the places where the survey results were administered. Given the importance of the road network to navigation in the Denver area, clear line symbols representing the main motorways are prominently included.

The map scale, as it appears in its printed form, is 1:8750,000. The width of each road is about .02 inches, wide enough to be legible. Since one unit of map distance represents 850,000 units of that distance in the real world and each road is .02 inches on the map, the motorways appear to be 17,500 inches wide (since .02 * 850,000 = 17,500), or about 1450 feet (circa 440 meters) wide! In the USA, highway lanes are around 12 feet wide, and so, in reality, most of these roads are not much wider than 70 or 80 feet (circa 25 meters). Clearly, the map symbols are a massive exaggeration of the width of roads.

The reason for this sort of manipulation should be obvious. If the roads were printed to scale, the largest roads would appear to be about one-thousandth of an inch on the

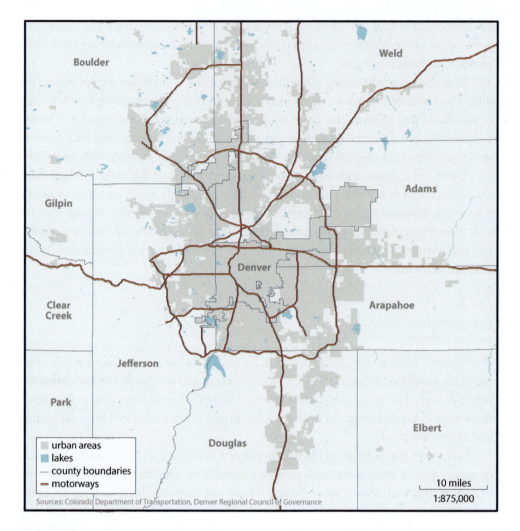

FIGURE 3.7
Example of a map with roads for reference.

map itself, about a third of the width of a strand of hair, which would be practically illegible. For maps to make any sense at all to us humans, they must ultimately perform what can seem like rather ridiculous manipulations of reality—or, as Monmonier might put it, they must "lie."

The topic of how to manipulate symbols on maps to make them more legible has a rich history of scholarship and is captured by the term **cartographic generalization**. Generalization is the graphical manipulation of symbols on a map to make them legible at the map's target scale. While generalization has been a part of cartographers' essential toolbox since the first map was made, there has been a great deal of scholarly research on the topic since the 1960s (e.g., Douglas and Peuker 1973, McMaster and Shea 1992) due to the need to program computer software to perform **automated generalization**, computer procedures that can perform generalization following strict operational rules.

Another definition of cartographic generalization that is driven by experiences in modern computer cartography is "a means for deriving smaller-scale maps from larger-scale maps and is a critical technique for the revision (updating) of smaller-scale maps from newly updated larger-scale maps" (Li 2007, 1743). *Google Maps* (https://www.google.com/maps/) provides a good example of effective automated generalization; if you zoom in and out of an area, you should be able to see changes to the features that accommodate making the map legible at a variety of scales. As you change the scale, some of the labels and features appear or disappear and the lines become more precise or smoother. The concept of generalization has proven challenging for cartographers to formally define because it has been difficult to construct clear and specific guidelines to govern the process, which often must rely on subjective human decisions. Drawing from this work, however, a basic set of generalization operators has emerged, specific techniques that cartographers (or the computers assisting them) can perform to generalize maps.

Before discussing specific generalization operators, however, it is important to distinguish between the concepts of *generalization* and *selection*, terms that are used inconsistently in professional and academic literature. **Selection** refers to the removal or inclusion of a set of features to make a map legible, which must occur on any map. While a map may include all the known instances of a class of features, the cartographer must carefully choose which facets of the geography to include. The map in Figure 3.7, for example, does not show buildings, forests, wildlife, or countless other geographic features. Critically, I only included major highways and excluded most of the rest of the roads because including all of them would render the map practically illegible. The map would not serve its goal well if too much other information was included.

Often a great deal of selection must be performed when mapping specific features. A good example of selection comes from the United States National Hydrography data set, which contains geospatial data on streams and rivers in the United States. When tasked with presenting a map of all the hydrological features in the USA, there are far too many rivers and streams to make a useful or appealing map (Figure 3.8).

In order to deal with this problem, multiple methods have been proposed to develop objective guidelines to pick out, or "select" the key waterways to efficiently produce a legible map. In this example, the authors drew from the precipitation regimes of the watersheds to drive how many rivers were shown for each region (Tinker et al. 2013). The resulting map in Figure 3.9 is a much better cartographic product with greatly improved legibility, and it still provides a fair and reasonable representation of the hydrography in the UA.

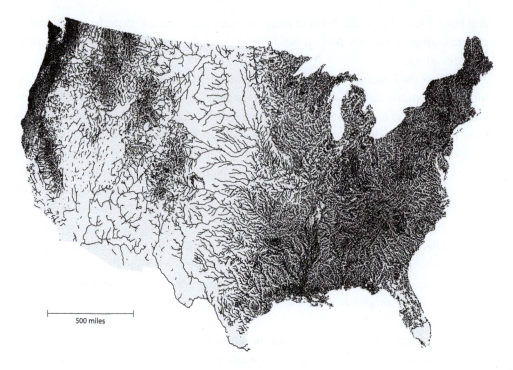

500 miles

FIGURE 3.8
A map from the US National Hydrography data set that does not include any selection.
Source: Tinker et al. 2013.

Once the features to be mapped are selected, generalizing the symbols is often the next step. The key underlying goals are to make the map legible, clear, and aesthetically appealing. Scholars have identified several conditions that require generalization in automated contexts (e.g., McMaster and Shea 1992, Slocum et al. 2009), so that computers can efficiently identify the need for generalization. **Congestion** is when too many features appear in too small a space, such as in Figure 3.8. Similarly, **coalescence** occurs when features are so close that they appear to touch, when they are, in fact, distinct objects.

Conflict is a special circumstance in which, due to some form of generalization, the symbols communicate an idea that is logically inconsistent with the rest of the map. McMaster and Shea (1992) give the example of a bridge that appears on a map, but the river it traverses is no longer visible due to selection, producing a puzzling situation in which the bridge appears to traverse nothing. **Imperceptibility** is the idea that a feature is too small to be visible to human eyes. Imperceptibility was the problem in the map showing highways above; the roads had to be augmented to "larger than life" size so that readers could see them.

FIGURE 3.9
A map from the US National Hydrography data set with automated selection.
Source: Tinker et al. 2013.

Related to these issues are some ideas about when and where to generalize. For instance, **complication** is the idea that the generalization process depends upon conditions in different areas of the map. In the aforementioned example of the hydrography map, the authors changed the threshold for how many waterways to show based on the precipitation regime (Tinker et al. 2013). A stream that appears in an area with lots of rainfall might be "selected out" in a drier area. Complication leads to **inconsistency,** referring to the idea that generalization decisions are not consistently applied across the map.

Generalization operators

In the following section, I will review common generalization operators (adapted from McMaster and Shea 1992, Slocum et al. 2009). While many of the following are more pertinent for automated than for traditional forms of cartography, the concepts and ideas are useful as you contemplate the ways in which maps "lie" for the sake of effective communication. All forms of generalization come at the cost of some accuracy, and so it is often a matter of achieving the right balance. As you review these operators, you can also consider how to use the methods to make your own maps more readable. A graphical summary of generalization operators is provided in Figure 3.10.

Simplification

Simplification is the process of selectively reducing the number of points in a line to represent an object. Until around the 1980s, when digital cartography was a nascent field, spatial data consumed a cumbersome amount of digital storage, and so it was important to devise the most efficient way possible to store it. It takes 11 points to define the line at the top of Figure 3.10. After running a simplification procedure on the line, a similar-looking line can be displayed with about half as many points. Particularly when the scale is reduced (when the map is "zoomed out"), such high precision is not necessary and may clutter the map. The simplified lines can be easier to view at the smaller scale without unnecessary detail.

Smoothing

Smoothing is distinct from simplification because it does not rely on removing points from the line, but rather on reducing the "angularity" of them to form a smoother, more flowing representation. Many modern graphical and GIS applications can effectively achieve line smoothing. While the points that make up a digital representation of a river might be very accurate, rivers do not come in the form of jagged lines. When people view a river on a map, smoothed lines are more readily interpreted as rivers and generally more visually appealing.

Exaggeration

Exaggeration, the amplification of a specific feature or object, is a common form of generalization because it is often critical to legibility. In the example provided in Figure 3.10, the map shows a bay; the peninsulas that form the mouth of the bay appear quite close. The critical feature of a bay is its access to an ocean or a larger body of water, an important idea the map should communicate. The mouth can be widened on the map to clearly preserve that message. Other common examples of exaggeration include enlarging the point symbols of cities on a map or increasing the width of lines that represent rivers or roads.

Refinement

Refinement is the idea that some specific portions of a network are shown to provide the reader with an idea of the characteristic and nature of the phenomenon being

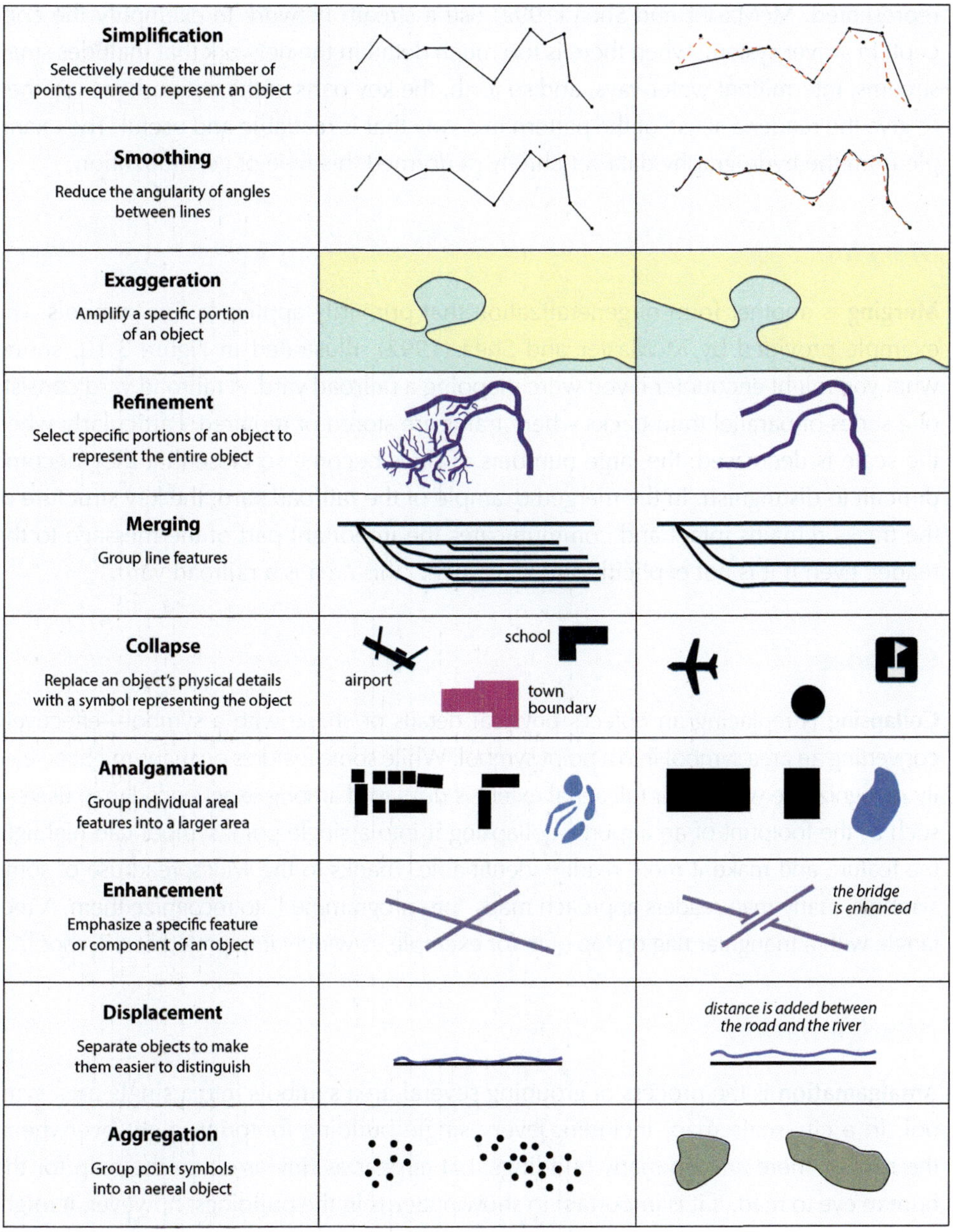

Simplification Selectively reduce the number of points required to represent an object		
Smoothing Reduce the angularity of angles between lines		
Exaggeration Amplify a specific portion of an object		
Refinement Select specific portions of an object to represent the entire object		
Merging Group line features		
Collapse Replace an object's physical details with a symbol representing the object	airport · school · town boundary	
Amalgamation Group individual areal features into a larger area		
Enhancement Emphasize a specific feature or component of an object		the bridge is enhanced
Displacement Separate objects to make them easier to distinguish		distance is added between the road and the river
Aggregation Group point symbols into an aereal object		

FIGURE 3.10
Types of spatial operators in generalization.

Source: Based on McMaster and Shea (1992).

represented. McMaster and Shea (1992) use a stream network to exemplify the concept. In a river system, when there is too much detail in the network that includes small streams, intermittent waterways, and so forth, the key parts of the network are retained to give the reader a sense of the pattern in a way that is readable and useful. The example from the hydrography data set chiefly performed this type of generalization.

Merging

Merging is another form of generalization that primarily applies to line symbols. The example provided by McMaster and Shea (1992), illustrated in Figure 3.10, shows what you might encounter if you were mapping a railroad yard. A railroad yard consists of a series of parallel train tracks where trains are stored or repaired. Particularly when the scale is decreased, the large numbers of lines become so close that they become difficult to distinguish. In the merged example of the railroad yard, the key structure of the tracks remains intact and communicates the important part of the message to the reader, even if it is not explicitly stated as such: "this area is a railroad yard."

Collapse

Collapsing is replacing an object's physical details or shape with a symbol—effectively converting an area symbol into a point symbol. While some features on maps might be easily recognizable when their full aerial extent is displayed among experienced map users—such as the footprint of an airport—collapsing it into a single point symbol can highlight the feature and make it more readily identifiable. Thanks to the widespread use of some symbols, many map readers approach maps "pre-programmed" to recognize them. A rectangle with a triangular flag on top of it, for example, is widely interpreted as a "school."

Amalgamation

Amalgamation is the process of grouping several area symbols into a single area symbol. In a city-scale map, including every single building footprint might overwhelm the reader; there are too many buildings that appear as tiny areas on the map for the human eye to read. If it is important to show patterns in the buildings, however, it might make more sense to group multiple sets of buildings in proximity to one another into a single area. Once again, some accuracy or "truth" is sacrificed in the name of communicating the key idea, rather than the specific details. The larger symbols communicate patterns in building footprints to the reader.

As with many forms of generalization, it is important to consider the impact on the map's accuracy in tandem with how people will interpret it. It might be worth including a small note to explain to the reader something along the lines of "building footprints show collections of buildings, rather than specific structures." Additional care must be taken when using some of these techniques because they can affect the underlying meaning of the information being displayed. If distinguishing between different structures is important given the intent of the map, then it may be worth considering other techniques or changes to the map (such as increasing its scale).

Enhancement

Enhancement is putting emphasis on a specific feature or component of an object, often with a particular objective in mind. In a map of roads, for example, you might use enhancement to indicate places where a road passes over another via a bridge. A bridge is extremely important to distinguish from a crossroad, for instance, if the map ends up being used to help people navigate through the road network.

Displacement

When two or more features—point, line, or area—are so close that they appear to touch or coincide, performing **displacement** is appropriate. Displacement moves the objects on the map further apart so that they appear as clearly distinct entities. A road that runs right along the edge of the river on a map with a scale of 1:24,000 practically appears to run on top of the river, which loses the important message: that there are two distinct entities on the map. Displacement sacrifices a bit of accuracy to ensure that the road and river are presented as distinct elements.

Aggregation

Aggregation is yet another example of grouping, but the operation groups multiple point locations into a single area symbol. This type of exaggeration gets at the fundamental cartographic decision of what symbol to use. If the scale is appropriately small and the purpose makes sense, such as a large map of a golf course, it might make sense to map individual trees because golf players probably care about the location of specific trees. It would make a lot less sense to map specific tree locations on a map of an entire city (even if you have the data). A much more cartographically sound approach would be to shade areas of the city to communicate where "wooded areas" or "forests" are located.

Summary

The goal of this chapter was to equip you with the essential concepts behind some of the many cartographic decisions that you must make when you produce a map. These decisions include what layout elements to include or omit, how to symbolize the data you want to convey graphically, and how much to alter or generalize the map to make it legible (sometimes with the price of accuracy).

The decisions you make about map elements, symbols, and generalization should be informed by the basic questions driving cartographic decisions: the map's purpose, audience, and medium. Once you have thought about those decisions, the next step is to consider decisions to designing the map itself. What exactly, then, makes a map with good *design*? How do you use map elements and features to produce a map that is both legible and appealing? If you were asked to produce a map, what are the guiding principles for designing one? The next chapter discusses the principles of good cartographic design, emphasizing both the science and art of cartography.

Discussion questions

1. The beginning of the chapter discusses the key questions behind map design: the audience, intent, and medium of the map. Suppose that you were tasked with building a map of COVID-19 infection rates for a general audience? How would the anticipated audience affect your map decisions? What would you change in the map if it were intended for a team of social scientists? What would you change if the map were intended for infectious disease specialists?

2. Using the example of building a map to show COVID-19 infection rates (or another example of a mapping goal), what visual variables would you use and why?

3. Examine the maps in the "Maps from the Wild" gallery. Do you agree with the map elements the authors included? Would adding map elements to any of the maps improve the map? Are there map elements that you would recommend removing?

4. Drawing from examples in the map gallery or other maps, which visual variables do you believe are particularly effective at communicating patterns in data? Would you change any of the visual symbology?

5. The foundational need to perform generalization and selection on maps often produces a tension: on the one hand, maps are not effective if they are not legible, and on the other hand, the inclusion of additional detail and accuracy can improve the information in the map. What principles can you devise to guide you through striking this balance; when is it okay to generalize and in what circumstances do the "lies" behind generalization compromise the integrity of the map?

References

Bertin, J. 2011. *Semiology of Graphics: Diagrams, Networks, Maps*. Redlands, CA: ESRI Press.

Bertin, J., and M. Barbut. 1967. *Sémiologie Graphique. Les Diagrammes, les Réseaux, les Cartes*. Paris; La Haye; Paris: Mouton; Gauthier-Villars.

Douglas, D.H., and T.K. Peuker. 1973. "Algorithms for the Reduction of the Number of Points Required to Represent a Digitized Line or Its Caricature." *Cartographica: The International Journal for Geographic Information and Geovisualization* 10 (2):112–122. doi: 10.3138/fm57–6770-u75u-7727.

Li, Z.L. 2007. "Digital Map Generalization at the Age of Enlightenment: A Review of the First Forty Years." *Cartographic Journal* 44 (1):80–93. doi: 10.1179/000870407x173913.

McMaster, R.B., and K.S. Shea. 1992. *Generalization in Digital Cartography*. Washington, DC: Association of American Geographers.

Monmonier M. 2018. *How to Lie with Maps, Third Edition*. Chicago, IL: University of Chicago Press.

Muehlenaus, I. 2014. *Web Cartography: Map Design for Interactive and Mobile Devices*. Boca Raton, FL: CRC Press.

Peterson, G. 2014. *GIS Cartography: A Guide to Effective Map Design, Second Edition*. Boca Raton, FL: CRC Press.

Slocum, T.A., R.B. McMaster, F.C. Kessler, and H.H. Howard. 2009. *Thematic Cartography and Geovisualization*. Upper Saddle River, NJ: Pearson Prentice Hall.

Tinker, M., P. Anthamatten, J. Simley, and M.P. Finn. 2013. "A Method to Generalize Stream Flowlines in Small-Scale Maps by a Variable Flow-based Pruning Threshold." *Cartography and Geographic Information Science* 40 (5):444–457. doi: 10.1080/15230406.2013.801701.

4 | Cartographic design

Recall the fundamental questions you should think about every time you embark on making a map: namely, the (1) purpose, (2) audience, and (3) medium of the map. What do these questions mean in terms of the actual design of a map? The principles of design often speak to the tensions in cartography between its practice as both an art and a science. While there are straightforward rules and conventions that should nearly always be followed in the practice of cartography, truly excellent cartography results from the ability to design a map that is both an effective vehicle of communication and visually appealing, given its specific purpose, audience, and medium.

The principles of design discussed in this chapter can serve as a helpful guide as you learn about the key skills in the *art* of cartographic practice. Students with an art or design background may begin learning cartography with some applicable and useful experience—giving them a knack for layout design—while others will need a lot of practice. With enough experience making maps—often after continually consulting with your map readers to learn about their view of the map and their reactions to it— the process of design can eventually become second nature. The principles discussed in this chapter are often related. Some of these principles produce tensions, requiring you to achieve some balance between competing concepts. All maps, however, should strive to achieve the following three goals:

1. The map should communicate relevant data clearly and efficiently.

2. The map should be structured in a way that guides the reader to its purpose.

3. The map should have integrity; it should present the data as honestly as possible.

I begin this chapter by discussing the two perspectives on design: the classical principles that have long appeared in cartography textbooks and a more recent body of ideas introduced by data visualization scholar Eduard Tufte. The drivers of much of the change we have seen in cartography over the last several decades are from changes in the technology that affects the cartographic media in which we may work. In the final part of the chapter, I discuss the historical and contemporary context of cartographic media and how they influence design considerations.

Classic principles of design

While the culture around cartographic design and the technology available to support the production and presentation of maps have changed dramatically and continue to evolve, some ideas have stood the test of time. I refer to these initial concepts as the "classic" principles of design because of their broad applicability to map-making and their long-standing relevance.

A good practice for improving your ability to apply these principles is to find examples of maps that violate them. Pay attention to maps that you believe are poorly designed and try to figure out what the problems are. Rather than reproduce existing poor cartography here, I have produced a rather exaggerated example of some bad cartography with a simple map that caricatures the problems I often see among novice cartographers. Figure 4.1 shows a map of the percentage of users by country who had access to the Internet in 2016, according to the World Bank. Figure 4.2 shows the improved(!) version that addresses the principles discussed here. Both maps have the same projection and use the same data, which have been categorized into quintiles

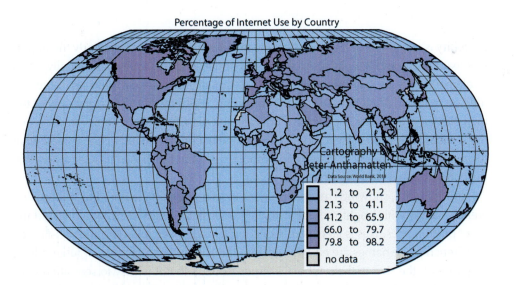

FIGURE 4.1
A map of Internet users per capita that lacks legibility, good visual contrast, figure-ground, a clear visual hierarchy, and balance.

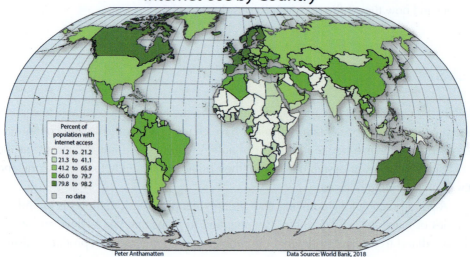

Internet Use by Country

Peter Anthamatten   Data Source: World Bank, 2018

FIGURE 4.2

A map of Internet users per capita with improved legibility, good visual contrast, figure-ground, a clear visual hierarchy, and balance.

(five groups containing an equal number of countries). The only differences between the two maps are derived from layout and design choices.

Legibility

Legibility is exactly what it sounds like; the features on the map, particularly the important features, should be readable and easy to understand. Legibility traditionally refers to elements with text; if the text has too little contrast or is too small to read, there is little point in including it on the map. In mapping, legibility can also refer to any of the map elements, including the map symbols and the map itself. The features on reference maps should not be so close to one another that it becomes difficult to understand what they represent. In thematic maps, the symbology should be chosen in such a way that the patterns in the data are clear.

In Figure A.2, showing French toponyms in Minnesota, the author has symbolized French names with bright red color, making the pattern easy to view alongside the other place names. This is an example of good legibility because both the non-French and French place names are easy to see. Similarly, in the map of cases of tuberculosis (Figure A.8), the size of the symbols is varied enough that it is easy to see differences in the patterns. The other maps in the map gallery generally exhibit good legibility by using graphical symbols that are easy to distinguish as well as text that is not too small.

In the bad map example presented here (Figure 4.1), legibility is a problem with text (particularly in the data source and authorship, located over the Indian Ocean) that interferes with the lines of the graticule; as I will discuss in Chapter 6 on typography, a key rule is to avoid letting symbols (especially lines) interfere with text. Legibility is also hindered by the way the data are presented with different shades of blue. This is generally not a bad choice among the visual variables—since the map is about quantitative data and the map appropriately uses lightness and saturation—but it is difficult to distinguish to which category specific countries belong to.

Visual contrast

Visual contrast refers to how the map features, elements, and symbols contrast with their backgrounds, other elements, and the layout in general. Visual contrast is particularly important for the map itself. Good contrast employs colors that yield a clear and crisp distinction between the map and the rest of the layout (or similarly, between the important and less important parts of the map). The map of solar power potential in Missouri (Figure A.3) employs visual contrast effectively; the state of Missouri contains orange and red symbols to convey the data, which contrasts sharply with the surrounding states, symbolized with light gray color.

Visual contrast is poor in the map example provided in Figure 4.1. The purpose of this map—the critical part that should be clearly communicated—is the information about the rate of Internet use by country. The lack of visual contrast causes the data to appear to blend in with the rest of the map. The countries themselves should be symbolized with colors that have enough contrast that readers can easily distinguish the data categories from one another. In the improved map, the colors are changed from a blue to a green spectrum to provide better contrast with the bluish ocean background. The ocean is "desaturated" somewhat in the improved version, bringing it closer to gray and providing better contrast to the main subject of the map.

Figure-ground

Figure-ground is a notion that the figure, or foreground, is visually separated from the background. Maps that effectively employ this principle appear to "pop out" from the layout, making it easy for readers to distinguish between the main map area and the background. Figure-ground is often related to the concept of contrast, since contrast is a good way to give the map some prominence. Clearly, the map should be one of the most prominent features, so the contrast should be designed in such a way that the *map*, rather than the background, attracts the eye and stands out. Using light-gray or near-light-gray colors for the background nudges other elements of the layout forward in the visual field. Examine the map of changes in protected area across the globe in

Figure A.12. The use of a prominent line symbol around the land area helps to separate the land area from the water. Other techniques include adding an outline or vignette around the mapped area and including a halo or shadow effect. The improved map in Figure 4.2 demonstrates the use of a drop shadow effect around the land area, which can dramatically improve the figure-ground. The subtle shadow produces a visual illusion that the countries are hovering above the rest of the map.

An additional technique, effective if appropriately executed, is to include a graticule as part of the background. The familiar lines of latitude and longitude are useful for helping astute readers orient themselves and can give important clues about the projection. Perhaps as importantly, the consistent pattern of lines can be made to look very much like a background, almost like lines between tiles on the floor in the background of a photograph, and can effectively contribute to building good figure-ground.

Visual hierarchy

A "hierarchy" refers to "an arrangement or classification of things according to relative importance or inclusiveness" (Oxford English Dictionary). The map layout should include a hierarchy of the elements that are arranged in order of importance to the reader. The concept of **visual hierarchy** refers to communicating the importance of different components of the map through graphical manipulation.

We can once again employ the language metaphor to think about the structure of maps. When you are writing a scientific report, for instance, it is often helpful to start a section or paragraph with the main idea or an overview of the main idea. In a map, the map reader should notice the map elements at the top of the visual hierarchy first. When you organize a map into a visual hierarchy, you can think of it in terms of *layers of meaning* and can manipulate the layout to guide the reader through those layers in a logical, sequenced progression.

Think about the first thing that people notice in a layout. *Size* conveys importance; the largest elements are the first things that map readers are likely to notice and read. *Position* also conveys importance; elements at the top and center normally fall higher in the visual hierarchy than elements at the bottom. *Color* can also affect to where the reader's attention is naturally drawn; bright or highly saturated colors stand out more than less saturated (closer to gray) colors.

Cartographers can manipulate the visual hierarchy not only to attract attention to the key layout elements but also to keep other elements on the same "visual plane" so as not to draw attention to any single one among a class of features. In the reference map of Europe (Figure A.4.), for example, the area symbols representing the countries and their labels are comparable in color, size, and style. The reader's attention is not drawn to any specific country—thus, an appropriate use of visual hierarchy in the reference map. In contrast, other elements in the layout occupy other positions in the visual hierarchy. The title is large compared to the rest of the text on the map and located at the center and top of the layout. The legend has a diminished position at the lower right side of the map; it is clearly legible and available for readers to consult, but does not compete aggressively for the reader's attention.

Visual hierarchy plays a similar role in thematic maps, but since the goal of a thematic map is to highlight patterns in data, the symbols for the data should be placed high in the visual hierarchy. In the map showing solar power potential in the state of Missouri (Figure A.3), the data are colored brightly in the center of the layout, and readers can hardly avoid focusing their attention on it. Other map elements include state boundaries, labels for the states around Missouri, major roads, and a few cities and towns within the state. These map features are included to put the state in context and to provide the reader with some orienting features, but they are not the main points of the map. The town labels are modestly sized, and the labels of surrounding states, while legible, almost blend in with the gray background.

Hopefully, you can identify several problems with the visual hierarchy in Figure 4.1. One of the most common visual hierarchy problems I observe on maps by novice cartographers is the placement of far too much emphasis on the legend or other map elements (the example I have provided is only a modest exaggeration of some of these maps I've seen!). The legend is critical for helping the reader to understand the map, but it is seldom the most important feature. Your goal should be to enable the reader to easily interpret the map, with a legend that is easy to find and read, but without visually prompting focus on the legend. In the bad map example, the size, color, and contrast of the legend make it the first thing many readers will see when they examine the map.

I generally like to place the title high in the visual hierarchy; its job is to convey the main idea or theme of the map rapidly and succinctly. The position of the title at the top and center of the layout in the improved version is the largest text on the page and should attract the reader's attention (Figure 4.2). However, if the map is well contextualized and leaves little question about the topic, the title may not be as important. On the other end of the visual hierarchy, the data source and author were moved to the bottom of the layout and given text just large enough to read. The data and author are worth reporting, but they are ancillary to the map's purpose.

There are several other subtle ways to adjust the visual hierarchy. Given that the data groups are generally geographically contiguous in the map examples, the country borders are worth including but are best represented with just-legible, thin lines. The lines used to represent the country borders in the unimproved map bog the layout down with the borders and almost drown out the colors that should represent the data. In the improved map, I have reduced the line thickness and made the borders semi-transparent to further diminish them. A similar principle applies to the graticule, which occupies too much visual weight in the first map.

Balance

Balance refers the distribution of map elements in the layout. A well-balanced layout uses the entire map space, conveys an impression of equilibrium, does not appear lopsided and uses the entire space effectively. One way to think about balance is in terms of visual weight and visual direction (Buckley 2012). In contrast to the concept of visual hierarchy, which refers to design elements that attract attention, **visual weight** refers to how much an element on the map dominates the view. A map with a lot of visual weight might appear too densely populated with objects, lacking "breathing room." A map that generally has too little visual weight, on the other hand, can become unbalanced by appearing too sparse and not using the layout space available. The unimproved map of Internet access has a lot of visual weight; the lines representing the borders and graticules are thick; and some of the map elements are packed tightly on part of the map along with the surrounding map features.

Visual direction refers to the "center of gravity" of the map, the point around which the weight of the layout objects appear to be centered. The goal is to produce a layout that has well-distributed visual weight and does not pull the readers' eyes too far from the center. Dent, Torguson, and Hodler (2008) discuss the distinction between the **geometric center**, the physical center of the map, and the **optical center**, where people focus their attention, often just above the physical center. Maps are poorly balanced when they convey a sense of being lopsided or biased to one side of the optical center, with too many elements appearing on a side of the map. Dent, Torguson, and Hodler (2008) also observe that the eye naturally starts at the top, left side of the layout and then moves to the bottom and right, a pattern which cartographers may take advantage of.

In the bad map example, all the ancillary map elements are on the right side of the map. Balance is easily improved by a few minor changes, such as moving the legend to part of the map where there is space without much going on, in the southern part of the Pacific Ocean. Centering the title at the top and placing the authorship and data information in smaller text on either side of the layout at the bottom of the map also improves the balance.

An important component of balance is the concept of **negative space**, also called **white space**. Negative space refers to the spaces and gaps between the elements. Including negative space is important for reducing the amount of work the reader's mind must perform to disambiguate the specific elements. In the example of the improved map, there is ample negative space around the legend—between the legend box and the coastlines of South America. The user does not have to spend much mental effort navigating around the legend. Too much negative space can make the map layout appear sparse and devoid of content, while too little negative space can make it appear heavy and cumbersome.

Summary of the classical principles of design

Map designs that adhere to these principles—legibility, visual hierarchy, contrast, figure-ground, and balance—are usually off to a good start. Naturally, these principles should be applied with careful consideration of the basic questions around the purpose and the intent of the map. If aesthetic appeal is important for a map—perhaps the map will appear in a magazine article or advertisement, for instance—then balance is probably critical. In other cases, where the map has a much more utilitarian purpose, such as a cursory exploration of data, usability and readability (legibility, contrast) may be much more important than visual appeal.

These and most other principles of design come with some inherent tensions which the map designer must negotiate. You can make an extremely legible map by ensuring that there is ample negative space around all the text, for instance, but that may come at the expense of balance. Ultimately, much of the process of design is about striking the appropriate balance between competing design guidelines to achieve some sense of visual appeal.

Eduard Tufte's principles for graphical excellence

I have found the ideas in one of Eduard Tufte's most notable works, titled *The Visual Display of Quantitative Information* (2001), to be extremely helpful, both for guiding map design and teaching cartography. Tufte is a professor at Yale University, with an established body of work on the presentation of informational graphics and data visualization. He discusses historical and contemporary developments in the design of data and informational graphics, and in so doing builds a case for the guiding principles behind *excellence* in data graphics.

While Tufte's book is not about maps per se, and only a portion of the examples of graphical communication in his book includes maps, the concepts he discusses apply well to cartography; both data graphics and maps are, after all, about communicating large amounts of often complex data in an efficient, graphical form. While the principles he articulates often seem to border on common sense, they are often neglected in popular media and academic literature, but are nonetheless worth learning and contemplating as you produce maps. In this and other work, Tufte also coined several useful terms that provide additional vocabulary for discussing graphical communication. The final component of this chapter reviews each of Tufte's five *Principles of Graphical Excellence* and applies them to cartography. As you read through this discussion, consider some of the specific and tangible implications that each could have for map design.

Graphical excellence is a well-designed presentation of interesting data—a matter of substance, statistics, and design

The first among the five principles Tufte presents is a good general overview of approaches to sound design. The central ideas here are that the design of the graphic should focus on the substance of the *information being communicated* rather than on some other aspect of the map or data graphic, employ statistical methods that honestly depict the data, and should be interesting and visually appealing.

Maps and data graphics should employ sound methodologies in the manipulation of presented data. Tufte considers decisions about how to convey statistics to be an important component of the design itself. In choropleth maps, for example, a common, but often misleading, practice is to report the *counts* rather than the *rates* of whatever phenomenon is being mapped.

A critical point attached to this principle is that the maps should drive the reader to focus on the data and the *substance* of the story being presented, "rather than about methodology, graphic design, the technology of graphic production, or something else" (Tufte 2001, 13). While it can be tempting to include an array of impressive-looking graphics to accompany maps—such as drop shadows, fancy fonts, or an ornate compass rose—the best maps tend to apply these tools and techniques judiciously to yield a simple and clear design.

Tufte strongly warns readers to avoid the inclusion of **chartjunk**, which includes "overly-busy grid lines and excess ticks, redundant representations of the simplest data, the debris of computer plotting, and many of the devices generating design variation" (Tufte 2001, 107). Chartjunk can show up in maps (as "**mapjunk**") through a variety of ways that include giving too much visual weight to grid lines or reference ticks, using unnecessary graphical ornamentation on the data themselves, or even using overbusy political borders or symbology on a base map. The key idea here is to consider the justification for the inclusion of every element on the map. If the substance of the data is clearly and elegantly presented, Tufte argues, the visual appeal of the graphic or map should stand on its own.

Graphical excellence consists of complex ideas communicated with clarity, precision, and efficiency

The power of data graphics and maps comes from their ability to assemble complex data into a form that humans can comprehend and explore. Given the complexity of data, it is not difficult to overwhelm the reader with information that undermines this power. Good design decisions clarify complexity.

An important point behind Tufte's second principle is that you should consider whether the map or data graphic is "interesting" or worth presenting in the first place.

Cartographers tend to exhibit some alacrity about including maps with their work. As you are working on a project, take a second to consider whether the map is a good use of space and your readers' attention. If the mapped data do not bring much to the project or the data could be more effectively presented as a table or some other form of data graphic, consider leaving the map out in favor of another form of communication. It can be easy to come up with things to map, but the production of the map should serve a clear function worth occupying the space.

Graphical excellence is that which gives the viewer the greatest number of ideas in the shortest time with the least ink in the smallest space

A goal of efficient graphic design is to maximize the **data-ink ratio**, which Tufte defines as the amount of *data ink*—ink dedicated to communicating variations in the data—compared to *non-data ink*—the rest of the ink on the map. Examples of non-data ink in aspatial data graphics include graphical objects such as the grid lines, lines connecting the points on a graph, or graphical adornment—anything that does not serve to communicate the data themselves and the patterns they reveal counts as "non-data ink."

Tufte argues that highly effective graphics *maximize* the data-ink ratio, and so he advocates for pruning out the components of any graphic that do not serve its primary purpose. In mapping, examples of non-data ink could include political borders or other base map information, the compass rose, neatlines, or grid lines. While all of these are often worth including on a map—to orient the reader, help the reader place the map data into context, or provide structure to the layout design—Tufte brings their utility into question.

The idea behind this principle can be applied to mapping by reviewing a few fundamental questions. As you consider "non-data" ink, think about whether removing those facets of the map would impede the reader's ability to effectively read the patterns of the data on the map. One technique Tufte suggests is to convey the non-data information by *erasing* the ink, thereby serving the same function by physically removing ink from the layout. If it is important to communicate the specific values on a bar graph, for example, he suggests erasing (rather than including) horizontal grid lines. An analogous technique in mapping is to indicate borders on a choropleth map by erasing the borders, rather than by printing them as lines (see Figure 4.3). For some mapping purposes, this technique can serve to highlight the data and diminish the visual weight of less-important features.

A distinct but related concept is **data density**, which Tufte defines as the "number of entries in a data matrix" divided by the "area of the data graphic" (Tufte 2001, 162). A data graphic that takes up an entire page but only communicates few original points

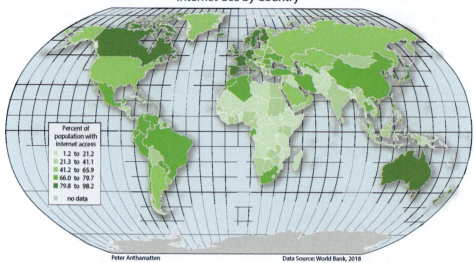

Internet Use by Country

Peter Anthamatten   Data Source: World Bank, 2018

FIGURE 4.3
An alternative version of the improved choropleth map of Internet use that reduces non-data ink.

of data is an example of low data density. Maps are generally data-rich by nature and are often worth taking up some layout space. However, if the map is a simple one that with low data density, you might consider whether you can present the data just as effectively using a smaller map that consumes less space. A choropleth map that shows voting results for a few states in the Northeastern United States, for example, is probably not worth an entire full-page layout. The same map, reduced to a quarter of its size and taking up much less space on the layout, can communicate the same data but with less space, improving the layout efficiency. The size reduction increases the data density because it reduces the denominator in Tufte's data density formula.

Graphical excellence is nearly always multivariate

The idea that graphical excellence is *multivariate* taps into Tufte's ideas about the enormous potential of graphical communication. He proclaims that a map from 1869 by Charles Minard, showing Napoleon's advance and retreat across Europe and into Russia, "may be the best statistical graphic ever drawn" (Figure 4.4; Tufte 2001, 40). What makes this map excellent is its elegant presentation of rich and complex data. This is essentially what we would call a flow map—the thickness of the line indicates the changing size of Napoleon's army over the course of its 1812–1813 campaign into Russia. The color of the line shows whether the army is in advance or retreat: the light brown line shows its advance, and the black line shows its retreat. Key battles are

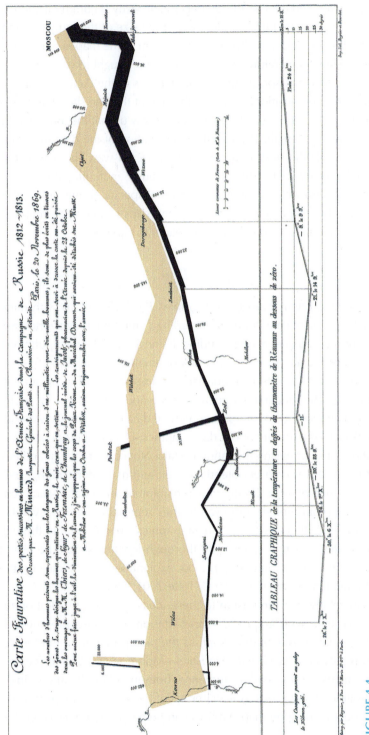

FIGURE 4.4

Charles Minard's map of the advance and retreat of Napoleon's army in 1812–1813.

marked on the map, and the loss of numbers by the army is communicated by steps in the line; the battles resulted in a sudden loss of the size of the army. The temperature during the army's retreat is also shown at the bottom of the map layout, demonstrating the role that the low temperatures played as the army sought to return to France.

What makes this map spectacular is the richness of the data and story that it tells through effective symbolization. Compare the size of the line shortly after it advanced from France, to its size upon its return—the data show us that the army was but a mere remnant of its initial strength. The visual representation of the data that relies on line width and color tells the story in a *clear* and *impactful* way. The map is multivariate because it includes several bits of data simultaneously, including the size of the army, the location of the army at key battles, the dates at specific points during the advance and retreat, and the temperature. Truly excellent maps can compile multiple types of data to tell a clear, data-driven story for the reader to explore.

Graphical excellence requires telling the truth about data

Tufte's final point is often heard in cartography. Producing maps with data integrity requires effort to present data honestly, including communicating shortcomings in the data. This point accompanies a long-standing theme around the manipulation and presentation of complex data. An insightful book originally published in 1954, titled *How to Lie with Statistics* (Huff 1954), emphasizes the importance of using and interpreting statistics critically. The fundamental idea of this work is that a lack of understanding of statistics and an inability to interpret numerical data open pathways for deception.

Tufte makes a similar point about data graphics. Data graphics contain additional pitfalls for the misuse or misinterpretation of data. Not only are the quantitative underpinnings of the data a source of potential deceit and misunderstanding, but so is the graphical representation of it. The graphical manipulation of data—such representing data that are not proportional to the values they represent, can be severely misleading. He recommends that designers of data graphics strive for clarity through explicit labeling. He writes, "clear, detailed, and thorough labeling should be used to defeat graphical distortion and ambiguity. Write out explanations of the data on the graphic itself. Label important events in the data" (Tufte 2001, 56).

Maps add yet another dimension to these ideas because, in addition to presenting quantitative data with graphical representation, they also present *spatial* data, which carry special considerations (such as the map projection). In *How to Lie with Maps*—inspired by *How to Lie with Statistics*—Monmonier (2018) discusses many of the ways that maps can mislead its audience, both intentionally and unintentionally, through cartographic decisions. An awareness of the ways that maps distort reality as well as how people interpret map symbols can help cartographers to produce honest maps.

Cartographic projection is an example of a special property of maps that should be handled with care. Because all projections distort the spatial properties of a map, cartographers should think about how it will be viewed and understood by the readers, and how to present distortions in a way that does not mislead the reader. The next chapter discusses the topic of projections.

The map on the top left of the set of maps showing the 2016 presidential election tallies (Figure A.10) is a little misleading simply because it shows the winner of each county. Because the winning Republican candidate, Donald Trump, was most popular among rural voters, many of the counties he won—rural counties colored red on the map—do not carry many votes. The counties with much higher population in the cities and along the East Coast were far more likely to vote for the opposing candidate, Hilary Clinton. This map makes it appear as if there was overwhelming support for Trump because it uses area to represent territory, rather than voters, and buries opposing votes in places where Trump received only a marginal majority. Political operatives wishing to make the case that there was a strong majority favoring Trump in the election might be tempted to use a map like this in a misleading way. A simple and straightforward solution is to present the rate or proportion of votes for Donald Trump rather than the total counts of population in some way, as in the top-right map in Figure A.10.

Designing for the cartographic medium

The final among the fundamental questions that drive cartographic design decisions is *what is the medium*? If you or your collaborators produce a map as for a chapter in a printed book, for example, you can safely assume that it will appear in print, at least for part of its life. In other situations, you may build a map to be projected as a slide in a spoken presentation. In most cases, you can anticipate that the map will appear in multiple formats, such as a printed article that is simultaneously published online.

Cartographers now perform most of their map design and preparation work on computer screens. Since maps appear on a wide range of media with different sizes, careful thought about the size of the final form of the presentation is important. Knowledge about the final size of the map can drive decisions you need to make about all facets of the map design, including the size of the text, the form of the layout, how to symbolize data, and even how much data to include on the map.

Another important consideration is how much time map readers will have to view and contemplate your work, often driven by the medium and context of the map. If the map will appear in a slide presentation to highlight a pattern, demonstrate a point, or to provide support to a single idea, for instance, it is unlikely that the audience will have more than seconds or a couple of minutes to examine the map, all the while listening to a presenter. The map should include little extraneous detail, contain stark and crisp symbols, and use text sparingly; it needs to get the point across quickly and clearly. On the other hand, a printed map in a journal or book may get much more viewing time. Not only should the reader have a very good sense of how the map fits into the story, he or she may spend time examining the details of the map to examine patterns, evaluate the meaning of the map in the context of the report, or even to generate original ideas about the phenomenon being mapped. A well-designed map that appears in a printed form can effectively incorporate a great deal of information and detail.

Printed media (hardcopy maps)

Despite the recent changes to the nature of cartography in the digital era, most cartographers continue to design maps that will end up on printed media at some point in their life cycles. A lot of cartography designed for internal use will eventually appear as a hardcopy map, such as map for a report distributed during a presentation or in an internal memo. You might also design a map to be part of a scientific report for a scholarly or professional journal, as part of a book chapter, or perhaps for a newspaper or magazine.

The technology behind modern printing is complex, but it is worth learning about some of the history, basic concepts, and terms. Most professionals frequently use office printers, of which there are several forms and types. **Impact printers** function by impressing ink upon paper, similar to typewriters. Those of us who can remember life in the 1980s will recall dot-matrix printers, which use print heads containing up to 24 pins to impress a pattern on a sheet of paper in patterns over an inked ribbon. Most dot-matrix printers are not able to produce color prints. **Drum printers** and **chain printers** are essentially computerized typewriters that similarly make ink impressions with print characters. When computer mapping first became a reality, maps were often built from assemblages of standard characters, such as "*" and "/" (see Figure 4.5). While some offices these days still contain impact printers, they are no longer particularly relevant for work in cartography.

More useful for mapmaking are **non-impact printers**, which offer much greater flexibility for printing color images. While the specific technology used by printers depends on the company that produces them, **ink-jet printers** function by spraying an ink pattern on a sheet from small droplets controlled through electrical charges. Inkjet printers are currently the most common type of printers in home offices. The printers arrange the ink into tiny colored dots that form an image when viewed from a distance. The main disadvantage to using inkjet printers is that the ink may smear or bleed a little before it dries, but they can produce high-quality color images.

Laser printers, in contrast, function similarly to photocopiers. A laser light is reflected upon a light-sensitive surface to build an electrostatic charge on the drum in a precise pattern. The ink is then transferred to the printed page with hot rollers, in the

```
+----------------+---------------------------------------------------------+
| ***      **    |                                                         |
|****  ******** |            ARCTIC SEA       Nordkapp                    |
|  *****ICELAND** |                           ////                          |
|  R*********** |                   //   //////////+++++                  |
|  ************* |                  /  //////////////+++++++++++          |
|    *******    |            L// //////////####/+++++++++++++++|
+----------------+            //  ////..##///####++++++++++++++|
|                              ////::::::##########+++++++++++++|
|                              ///.:::::::::##########++++    ++++|
|     ATLANTIC SEA            /////:::::::::::::#######++++++       |
|                             ////::::::::::::::###########+++++++      |
|   %                        /////:::::::::::::::  #######+++++++++++|
|  F%  %                     /////:::::::::::::::  #########+++++++++|
|                           ////////.:::::::::::  ###########+++++++++|
|                          //////////:::::::::::  ##############++++++++|
|                         ///////////:::::::::::  #####FINLAND####+++++++++|
| Capital cities:         ///NORWAY//::::::::::  #############+(Russia)+|
|                         //////////:::::::::::.  ##############++++++++++|
| C = Copenhagen          /////////O:::::::::::::  ##########++++   +++++++|
| H = Helsinki            ////////  ::::SWEDEN:::: #A  ###H       +++++++++++|
| O = Oslo                /////     :::::::::::S:        ++++++++++++++++|
| R = Reykjavik    NORTH           :::::::::::     +  +++++++++++++++++++|
| S = Stockholm     SEA     %      ::::::::: .:   ++ +++++(Estonia)++++++++|
|                          %%%     ::::::: :G       ++++++++++++++++++|
|         DENMARK -->   %%%%% %%% :::          ++  +++++++++++++++++++|
| Islands:              %%%%  %%%C    BALTIC +++++++++++++++++++++++|
|                       %%%      B%   SEA    ++++(Latvia)++++++++++++++|
| L = Lofoten Isl.      ++++                 ++++++++++++++++++++++++|
| B = Bornholm          +++++++    ++++++  +++++(Lithuania)+++++++++++++|
| F = Faroe Isl.  ++++  +++++++++++++++++++++++++++++++++++++++++++++|
| G = Gotland   +++++++++++++++++++++++ (Poland)++++++++++++++++++++++++|
| A = Aland    ++++++++++++ (Germany)++++++++++++++++++++++++++++++++++|
|              ++++++++++++++++++++++++++++++++++++++++++++++++++++++|
+---------------------------------------------------------------------+
```

FIGURE 4.5
An example of an ACSII map of Scandinavia (Lahelma and Olson 1994).

pattern left by the charge. Both inkjet and laser printers operate on the principles of traditional ink color combinations, and so color is produced by mixing patterns of CMYK (cyan, magenta, yellow, and black or "key") ink.

Plotter printers are capable of printing on paper using pens, originally produced for design and engineering applications. These printers utilize a vector-based printing model based on points and lines. While vector-based printing has become uncommon, large format printers, capable of printing on a roll of paper several feet wide, are often called "plotters," or merely "wide-format printers." Modern plotters typically employ inkjet technology. Because of the size in which they can print, plotters can be useful for printing conference posters or large maps.

An important feature of printers is their **print resolution**, which is measured in **dots per inch (dpi)**. The dpi refers to the graphical acuity of the printer; the higher the dpi, the more detailed and precise a print can appear. Most inkjet printers can produce prints between 300 and 730 dpi. Expensive laserjet printers can produce graphics with as high as 2,400 dpi resolution; however, lower resolutions are generally sufficient for map work. Journal and book publishers generally request that you send a high-resolution image for the map, usually in the range of 600–1,200 dpi. As a rule, you should try to make graphical files of any map you wish to print at least 300 dpi.

One of the key challenges of working with printed media is that the image you develop on a computer screen appears different from the printed version. I always recommend that you print hard copies of your maps to critically examine the map in its intended destination format. Printed colors appear much darker than they do on a computer monitor, particularly when they are printed with inkjet printers. Pay special attention to the color's value or lightness; a basic solution to adjust your display image for printing is to raise the value of the colors of the map by 10 or 20 points. Even if you are preparing a large map for a poster, it is nearly always worth the added expense of printing a semifinal version of your work so that you can make adjustments. Finally, if you encounter a situation in which you anticipate that the map will be copied multiple times (such as in a school setting)—perhaps you expect that users will even make copies of copies—mimic the process by making a few copies of copies to check the quality of the map. A few design tweaks, such as changing text to bold font or making the elements larger, can go a long way in preparing the map for its life cycle.

If the map will appear in print, determine the dimensions of the final, printed product. Suppose you wished to produce a map of the entire globe, with a width about twice the length of its height, using a traditional northern orientation on a common print-sized sheet of paper. If you determine that the map will appear on a single page with dimensions of 8.5 inches wide by 11 inches long in a portrait orientation (with the longer side vertically oriented), for instance, you can calculate the final print size. You should generally plan to give the layout an inch or so of space around the margins of the printed product, and so you can plan for your map layout to be a maximum of 6 or 6.5 inches wide and 9 or 9.5 inches long. The final map will therefore have a horizontal extent of about six inches and a vertical extent of around three inches (Figure 4.6).

Knowing with any certainty about the final format of your map is a bit of a luxury, since most cartography projects demand a great deal of flexibility. Geospatial and graphical applications enable users to specify the layout size, and you can adjust the text size accordingly. For printed maps, the smallest text should be reserved for map elements low in the visual hierarchy and should not go lower than about six points at the target print or display size. Large text elements, such as the title, can operate

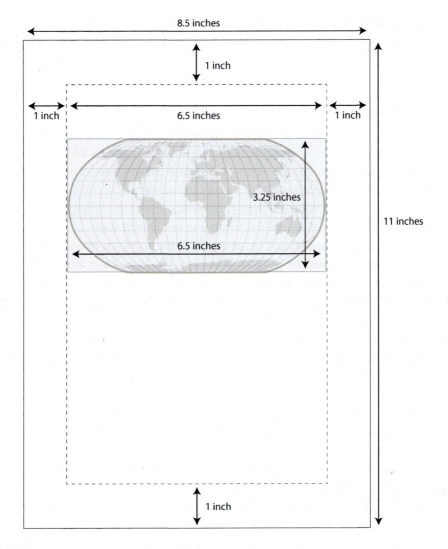

FIGURE 4.6
An example of a layout on a standard sheet of paper using a world map in a Robinson projection.

in the range of around 20 points, and other elements that should be clearly legible for a broad range of readers should be around 10 or 12 points. The very best practice is to print copies of the map at its target size so you and others can critically review the map.

If other people are responsible for preparing the final version of your work, bear in mind that the dimensions of the final form of the map may change. I have produced maps for print, believing that I understood how it would appear, only to find that the size of the final printed version was substantially reduced, which rendered much of the map illegible! Ultimately, the best process is to collaborate with everyone involved in the process at all stages of production.

Conference posters and large-format layouts

If the map layout is destined for a printed poster, such as in a large-format advertisement or a poster for a conference presentation, you should start by thinking about how the size of the layout space and how the map will be viewed. Some conference posters can be as large as two meters wide by a meter in height.

A good method for conceptualizing the design for printed posters is in terms of "layers of viewing." Passers-by should be able to easily read some facets of the layout from a distance; thus, text should be designed to be legible at as much as 20 or 30 feet, and so the text should be quite large indeed. The largest text, such as the title of the poster, could be around 80 or 100 points. Following that, you might include a phrase or sentence-long summary in smaller text, and then perhaps the author and the supporting institution. These elements should be designed to capture attention, attract interest, and draw viewers closer to your work.

After you have added the elements that convey the main ideas, you can add another layer for the interested viewer: envision a small group of people standing two or three feet from the poster, as you charmingly engage them about the meaning and importance of your work. The specific details of the map and layout, such as the legend, scale bar, and body text, should generally be sized around 24 points.

In the final layer of design, you are planning for the viewers you have successfully lured in. You can include some specific details and captions for readers who take the time to scrutinize your work carefully, at around 18 points. If the map is a central feature of the work, give it a large, prominent part of the main layout and ensure that the symbols effectively show the key patterns at a distance. Maps with lots of detail seldom work effectively in large-format prints, so try to stick to the main ideas and avoid including detail that is not critical to the key message. While it can be tempting to include paragraphs of text on a poster, few onlookers are likely to spend more than a few minutes examining your poster in most conference settings. A good rule of thumb is to keep the map and all the ancillary elements simple, clear, and succinct. The best posters also include plenty of negative space with lots breathing room in the layout.

The overall layout and design of conference posters should be well organized and balanced. Try to develop a layout that focuses around an attention-grabbing centerpiece (such as a key map) and carefully align the sections and body text around it. You can draw from lots of good online resources to help you with the design of conference posters, including websites with copious examples of good poster templates for a variety of software applications.

Digital media (display maps)

Even if you are planning to produce a map that will be printed, you can usually antic-ipate that your map will appear in a digital form such as in an electronic document, in a presentation, or on a web page. Digital maps comprise two key categories: **static maps,** which are little more than a computer image, and **dynamic maps**, with which the user can interact.

Raster images (images built from a grid containing cells of different colors) are generally best suited for static maps and are limited by the native resolution of the raster; if a user wishes to zoom in on a raster image, the pixels will occupy more of the viewing space and the image will become blocky or "pixelated." If you are not happy with the quality of a raster-based format, you can improve the quality by increasing the image resolution. Vector-based images, in contrast, are built from points and lines; the ability to change the scale in a vector-based image is limited only by the precision of the points used to encode it. Consequently, vector-based image formats are often preferable for digital display because they can better accommodate changes to the display size.

Slide presentations

If you are designing for an oral presentation fueled by an application such as *Power-Point* or *Prezi*, you may have little idea of how large the map will ultimately appear. If the presentation is delivered to a small office or workgroup, it may appear on a large television monitor several feet wide or on an individual's laptop monitor. In a lecture hall for a large class at a university, on the other hand, the map layout may show up on a screen several meters wide for an audience that consists of a few hundred students.

A good general rule of thumb for presentations is to avoid using small text; keep text at least 18 or 20 points. Depending on factors such as the power of the projection equipment, the lighting, the quality of the screen onto which the image is projected, and the size of the room, the contrast is likely to be diminished compared to computer monitors and printed maps. Consequently, it is good practice to ensure that there is high contrast—you might even exaggerate the contrast a bit for the text and important facets of the map.

Due to potential problems with contrast, it is also good practice to avoid includ-ing detail in your map that requires readers to make fine distinctions between colors, such as **hillshading** (subtle elevation shading to better convey the shape of the land). Pair down the map elements that are not necessary and give the remaining map ele-ments bit more prominence than you might otherwise do. While you should always avoid giving some ancillary map elements *too* much prominence (such as the legend, scale bar, or compass rose), these can be essential for the legibility of the map, and so you can adjust their size to ensure legibility across a broad range of presentation contexts.

Always remember that you can seldom count on readers to have a lot of time to understand the main purpose of the map and extract the main idea. Maps intended for live presentation should be designed to be readily understood through a simple design with bold, high-contrast colors and symbols.

Web and mobile cartography

Discussing cartographic media in the modern context is difficult because maps can assume such a broad variety of forms and formats. The field of digital cartography has emerged only over the last twenty or thirty years and continually evolves. While modern technologies provide both new capabilities and constraints for maps, the fundamental principles that guide the construction of good maps nearly always apply.

Web maps can be distributed as stand-alone layouts, such as in a .pdf document or image of the map itself, or they can be fully incorporated into the design of the web page. Most modern web maps are directly incorporated into the page itself and often include functionality that can make the map interactive by giving the user the option to change the view of the map, toggle layers of data, or adjust the symbolization.

Mobile phones and other devices introduce a novel set of design considerations, addressed elsewhere in this book and covered extensively in other texts (see, for example, Muehlenhaus 2013). Many modern mobile phone applications are programmed using the **Application Programming Interface (API)**, which now enables the use of **dynamic text**. Dynamic text can change its size and style that accommodate the resolution and size of the screen it is viewed on. Modern GIS applications effectively support dynamic text by keeping the text at a consistent size, regardless of how the user manipulates the scale, and removing labels from the map when it becomes too crowded.

The key advice (this should sound familiar) is to anticipate the contexts in which the map will be viewed and think carefully about who will use the map, what the purpose is, and how it will be used. If you are designing maps that will be used in situations where Internet access is poor or where many of the users may not have access to the latest computer technology, try to ensure that the maps can be sufficiently viewed on older monitors. If the Internet speed is likely to be a problem for users, strive to encode the maps in well-compressed, low-data formats. If you are concerned about users needing access to high-quality versions of your maps with high-resolution imagery, you can consider including a "full resolution" quality image alternative as a link, to give users the choice.

Ian Muehlenhaus (2013) makes some important recommendations for designing interactive and mobile maps. **Slippy maps** that enable the user to pan around the interface can offer exceptional flexibility to the user, but places constraints on what you can do as the designer. Particularly when designing maps for smart phones and other mobile digital technology, remember that space is precious; you should pare the map symbology down to the most essential elements. Muehlenhaus also recommends that cartographers avoid using compartmentalized interfaces on maps destined for smart phones; there is simply not enough space to accommodate all the non-data "ink."

Given the diversity of formats in which digital maps can appear, it is nearly impossible to provide strict or consistent guidelines. At the time of this writing, mapping applications have been launched for use on smart watches—interactive maps not much larger than a square inch. The best solution is nearly always to learn as much as you can about the targeted format and take the time to test the work in that format. Solicit feedback from users whenever possible so that you can adjust the work. After you have user-tested your mapping work and it is ready to launch, perform some additional testing!

Summary

Applying the principles of design discussed above can assist you in your efforts to produce a map that is clear, well-structured, and honest, but you must learn to juggle competing facets of the design to yield a map that suitably meets the overarching goals of the work. You and your audience are bound to have strong preferences about map design (color choice, for example, seems to be a particularly popular topic for personal opinions). Always bear in mind that your or others' preferences may be a poor guide for good cartography. The best teacher of all may be your own experience. Get into the habit of discussing your map work with others; find out what other viewers like about your design to determine which practices best achieve the map's objectives and nudge your work toward excellence.

Discussion questions

1. In the discussion, the text mentions that the importance of the classic principles of design—legibility, contrast, figure-ground, visual hierarchy, and balance—is driven in part by the audience, purpose, and medium of the map. Which of these principles do you think might be worth sacrificing in favor of other design choices given a particular mapping goal? Describe some situations in which that may occur.
2. Visual hierarchy can be a difficult concept to practice well. Apart from adjusting the size, position, and contrast of map elements, what are some specific techniques you can employ to adjust the visual hierarchy of the map's elements?

3. Find an example of poor cartography on the Internet (not an especially difficult task!). What could be improved on the examples you found? Which of the principles of design need attention?

4. Tufte's principles of graphical excellence were developed with informational graphics in mind. Are any of the ideas more or less applicable to producing *cartographic* excellence than others?

5. Beyond what was discussed in the text, can you identify any *tensions* that can result from "overdoing" any of the design principles discussed? Can you identify any examples of maps that demonstrate the tensions?

References

Buckley, A. 2012. "Make Maps People Want to Look at." *Arc User* 5:46–51.

Dent, B.D., J. Torguson, and T.W. Hodler. 2008. *Cartography: Thematic Map Design*. New York: McGraw-Hill Higher Education.

Huff, D. 1954. *How to Lie with Statistics*. New York: Norton and Company.

Lehelma, A., and J. Olson. 1994. "The Nordic FAQ." Accessed July 8, 2020. http://www.lysator.liu.se/nordic/index.html

Monmonier, M. 2018. *How to Lie with Maps, Third Edition*. Chicago, IL: University of Chicago Press.

Muehlenhaus, I. 2013. *Web Cartography: Map Design for Interactive and Mobile Devices*. Boca Raton, FL: CRC Press.

Oxford English Dictionary. "Hierarchy." Accessed August 21, 2018. https://en.oxforddictionaries.com/definition/hierarchy.

Tufte, E.R. 2001. *The Visual Display of Quantitative Information*. Cheshire: Graphics Press.

5 | Coordinate systems and projections

Introduction

Recall from earlier in the text that the term "map" is derived from an old word for a flat "sheet" or "cloth." The two-dimensional nature of cartographic representations presents a long-standing and distinctive problem for cartographers because the reality that maps should represent—the surface of the earth—has three dimensions. The process of **projecting** a map refers to transforming the curved spheroid of the earth's surface into a two-dimensional surface suitable for representation on a map.

Thinking about how projections distort space on a two-dimensional representation of the earth's surface is an important part of the practice of cartography. For large-scale maps with small extents, a well-chosen projection can result in little distortion. For small-scale maps that cover a much larger area, such as a region or country, you must decide which properties of map projections—most critically *area* or *shape*—are the most important to preserve.

The goal of this chapter is to introduce you to the ideas and concepts of coordinate systems and projections, guide you through the principles that can help you select a projection for your cartographic work, and familiarize you with some of the common projections that you are likely to encounter throughout your mapping career.

Coordinate systems

Take a minute to think about how you would describe a location to someone who has no idea where you are. Perhaps you rely on existing points of reference, such as "Suffolk, UK." Where is "Suffolk," though? You can describe Suffolk in terms of other locations you expect people to know, perhaps by describing the location as "on the west coast of England" or "about 50 miles northeast of London." Space is boundless, and so ultimately *all descriptions of location are relative*. It is impossible to describe a location without evoking some other known location, even if that location is an arbitrarily defined grid, such as lines of latitude and longitude.

Consistent and accurate mapping relies on the systematic description of locations. Human civilizations have long struggled with devising a means of consistently and accurately describing locations, using different "anchors" against which to relate location. Initial efforts relied on observable astral bodies such as the stars and the sun. By measuring how far the sun appeared above the horizon, for example, ancient Phoenicians could determine how far north or south a location was.

A basic solution to describing location is the use of a **coordinate system**. A coordinate system has specified axes against which the locations of points in space can be described. A two-dimensional **Cartesian coordinate system** contains a horizontal x-axis and a vertical y-axis, against which all other points are located. In Figure 5.1, the point (2,4) of a coordinate pair shows a location on a Cartesian grid. The first number "2" indicates that the point is two units of distance to the right of the y-axis and the second number shows that the point is four points above the x-axis. The second point listed below is at (–3,–1), three points to the left of the y-axis and one point below the x-axis.

A Cartesian coordinate system is ideal for mapping because it is planar (it exists on a flat plane). **Planar coordinate systems** provide a means for describing any point, line or polygon in a consistent manner, providing all the benefits of traditional analytical geometry, leveraging the power of several branches of mathematics. Systematic and accurate mapping requires the locations of the axes and the units of distance to be clearly and unambiguously specified. Calculating distance and area is extremely

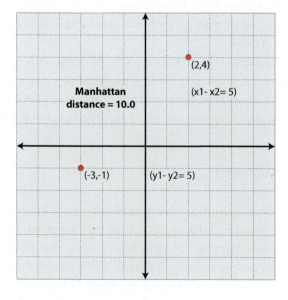

FIGURE 5.1
An example of a Cartesian coordinate grid system.

straightforward and a simple function in planar geometry, for example. If the planet were a flat, two-dimensional surface, the prospect of map-making would be very simple indeed.

A Cartesian coordinate system is not appropriate for a sphere, however. A line connecting two points on a sphere is not straight, but rather bends around the curved surface of the sphere over three-dimensional space; there are no parallel lines on the surface of a sphere. Specifying locations on a sphere requires a **spherical coordinate system**. Eratosthenes, a Greek scholar who lived about 2,300 years ago, is credited with the initial development of the **geographical coordinate system**—the spherical coordinate system most commonly used to describe locations on the earth.

Like all coordinate systems, geographical coordinate systems must begin with lines of reference against which to relate all other points. The very idea of a "line" is complicated because any line on the surface of a sphere is constantly curving. The closest a sphere can get to the idea of "straight line" is with a **great circle**—the line formed by any cut that divides the sphere into equal halves. Great circle lines have several useful properties. If you want to travel between two points on the earth's surface, the most direct route always follows a great circle, called a **great circle route**. Intersecting great circles on a sphere always form right angles. All other lines constructed by passing a plane through the sphere that does not cleave it into two equal halves are called **small circles.**

In the geographical coordinate system, the earth's **equator** plays the role of the x-axis, and the **prime meridian**—a great circle that passes through Greenwich, England—serves as the y-axis. Locations are defined in terms of angles from imaginary planes that pass through those great circles. The **latitude** is the angle and direction on the surface of the earth from the equatorial plane. The **longitude** is a similar measure against the prime meridian. Medina, Saudi Arabia, for example, is located 25 degrees north of the equatorial plane, and so its latitude can be described as "25°N." Its longitude is 40°E, which means that it is found at 40 degrees to the east of the plane formed by the prime meridian (Figure 5.2). The earth's lines of latitude are referred to as **parallels,** and the lines of longitude as **meridians**.

The conventional way to report geographic coordinates is to first list the latitude and then the longitude (e.g., 25°N, 40°E). Each degree is divided into 60 minutes, which are further divided into 60 seconds, called a **sexagesimal** system. The Eiffel Tower in Paris, France, for instance, is located at 48° 51′ 29″ N and 2° 17′ 41″ W. To facilitate use with computer applications, geographical coordinates are more commonly reported in a **decimal** system that uses negative coordinates to report locations south of the equator and west of the prime meridian. The decimalized coordinates for the Eiffel Tower are +48.858, +2.295. Santiago, Chile, located south of the equator and west of the prime meridian, at about 33° 30′ S, 70° 40″ W, then becomes −33.50, −70.67. Pay careful attention to the protocols used for coordinate systems in software. To better

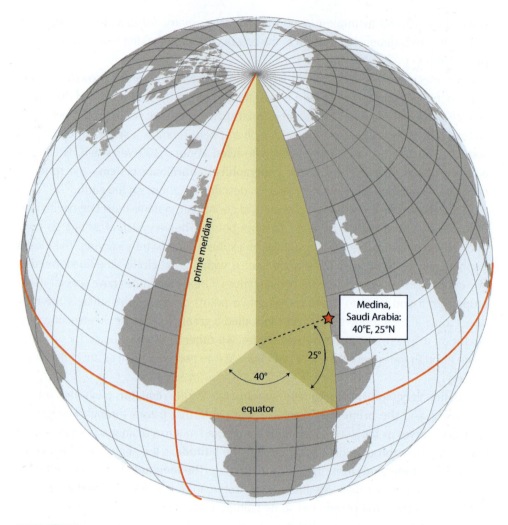

FIGURE 5.2

An illustration of the earth's geographic coordinate system, using Medina, Saudi Arabia as an example.

reflect common practice in coordinate grids, decimalized coordinate systems are usually reported with *longitude* first in geographic information and compute-mapping applications, since east-west looks like an x coordinate.

A complication from using a spherical coordinate system comes from the fact that the earth is not a perfect sphere. Due to the centripetal force from the rotation of the earth, it bulges at the equator. The true shape of the earth is an **oblate spheroid**, a sphere that is slightly flattened. The diameter of the earth is 12,714 kilometers from pole to pole, but it is 12,756 kilometers in diameter—an additional 42 kilometers—at the equator ([NASA] 2017). The flattening ratio is about 1:300, which means that the earth is about 1/300th shorter along its polar axis than along its equatorial atlas. You can imagine the shape of the earth to look roughly like a well-inflated basketball with someone sitting on it.

The bulge in the earth does not present a major problem for building a good geographic coordinate system; it is not too difficult to build a model of a coordinate system from an ellipsoid if the properties are consistent. The earth, however, is even more complicated. Its true surface has irregular depressions and bulges, driven by the composition of the earth's interior, which affect the gravitational pull of the earth's surface in different ways at different locations (Figure 5.3). The earth is, in fact, the shape of a lumpy **geoid**, defined as the shape of the surface of the earth if it were completely covered with water, essentially indicating where "true sea level" is at different places on the earth's surface (United States Geological Survey 2020). The study of the earth's shape is called **geodesy**.

Many mapping applications, such as navigation, surveying, and military operations, require describing locations with high accuracy. One requirement of any accurate coordinate system is that the reference points must be extremely clear and consistent. This leads to the question of how to construct a systematic coordinate system on an

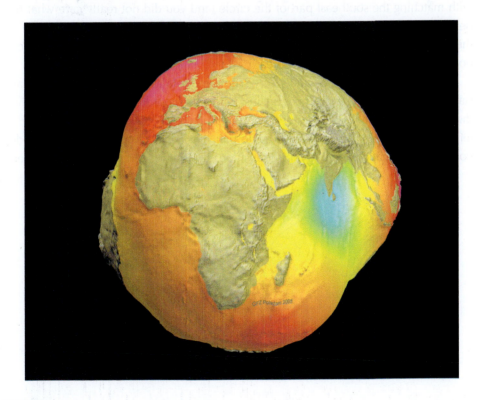

FIGURE 5.3
A highly exaggerated representation of the earth's geoidal surface (NASA).

inconsistent surface. The solution is to use a **geographic datum**, a reference ellipsoid modeled to provide the most accurate geographic coordinate system for the area being mapped. Most global coordinate systems, such as the one reported by most global positioning systems (GPS), use the **World Geodetic System of 1984,** normally referred to by its acronym, **WGS84**. This geographic datum serves worldwide applications because it does not distort any part of the earth to an extreme degree. Specific datums are defined to match a region of the world as closely as possible (to most closely match the bulges or depressions over the area being mapped). The US government, for example, generally uses the North American Datum of 1983 (NAD83) for maps of the United States, while the European Union recommends using the European Terrestrial Reference System 1989 (ETRS89) for maps of the European continent.

Figure 5.4 provides a hypothetical example of two different geographic datums used for different purposes. The slightly squat and lumpy gray circle represents the earth, bulges and all. If you were to produce an ellipse that most closely matches the perimeter of a circle, the red dotted line on the left might work the best; it does a reasonable job of representing the whole circle because the ellipse does not deviate too far from the actual perimeter at any point. If, however, you were chiefly concerned with matching the southeast part of the circle (and you did not really care what happens in other parts of it), you might draft out the green ellipse on the right. Notice that the green ellipse matches the southwest part of the circle nearly perfectly, but it does poorly over most of the rest of the area.

Geographic datums essentially serve the same role as the axes in a planar Cartesian coordinate grid, as references against which the coordinates are specified, giving them meaning. A latitude-longitude coordinate pair in WGS84 shows a slightly different location from the same coordinate pair that is based on the ETRS89. Modern computer applications enable rapid transitions between different coordinate systems,

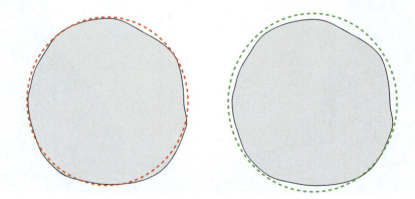

FIGURE 5.4
An illustration of different reference ellipsoids.

but it is generally advisable to minimize moving between geographic datums because it can lead to some distortion through rounding and other types of error. Choosing a geographic datum for a mapping project is usually driven by the projection, which are usually paired with specific projections. The precise coordinate system does not matter a great deal in small-scale mapping. However, if you are working with larger-scale maps in which some degree of accuracy is desired, it makes sense to work with a geographic datum that is most appropriate for the region of the world being mapped. When you draw data from formal government institutions, it is often a good idea to use the datum that is used as the standard for the part of the world you are mapping.

The latitude/longitude geographical coordinate system is well established and has a long history of use and application, so why not simply use it for mapping? Because maps traditionally appear on a flat surface, it is essential that geographic coordinates be translated into a planar system. It is not difficult to think of some of the challenges that emerge as a consequence. Geographic coordinates are reported in angles, which have different distances depending on where you are on the earth. If you traveled east along the equator, you would have to travel about 111 kilometers to traverse one degree of longitude. At the 45th northern parallel (45°N)—which passes through New York City and just north of Bordeaux, France—there are only about 79 kilometers between the lines of longitude. You can visualize this easily if you examine the graticule on a globe and can see the meridian converge at the North and South Poles; notice how the meridians converge at the North Pole in Figure 5.2. While it is possible to calculate distances from angular coordinates by drawing from principles of trigonometry, it is complex and cumbersome.

The various means of transforming coordinate systems from a *geographic*, angular system, to a *projected*, planar coordinate system affect the nature of distortion on maps, and so understanding projections is critical for cartographers. A classic technique for teaching students about the realities of transforming a sphere into a plane is with an orange. Try (or imagine trying) to place an orange peel on a flat surface. It is not possible to flatten the peel without tearing, stretching, or shearing it. Part of the job of the cartographer is to guide this projection process to produce maps that minimize distortion as much as possible and to choose which kind of distortion is the most appropriate for the mapping purpose at hand. The next section describes the characteristics of projections and provides some guidance for working with them when you make your own maps.

Projections

Imagine that the earth were a translucent globe with a graticule next to a flat sheet of paper. If you shine a bright light from the center of the earth, the shadow of the graticule would be projected on the paper. The term "projection" and much of the technical

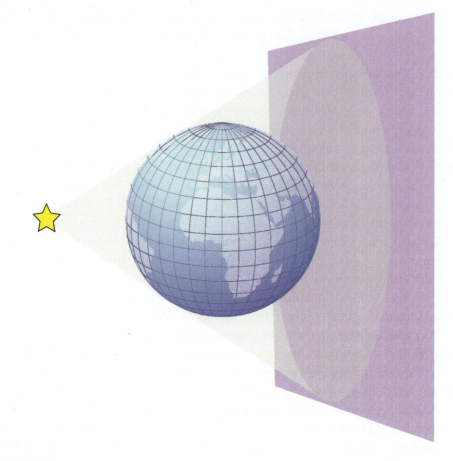

FIGURE 5.5
A simple illustration of the concept of projection.

terminology related to projections are rooted on this basic idea—the transformation of the surface of the earth's sphere (or, more precisely, its oblate spheroid!) to a flat surface through a mathematical process that simulates such a literal projection (see Figure 5.5).

All projections are based on a **developable surface**, a three-dimensional shape that can be flattened into a plane without any distortion or tearing. Turning back to the orange analogy: if you imagined that the orange were the earth and wanted to project it onto a sheet of paper, how could you contort a *sheet* in such a way so as not to tear or stretch it? There are three types of developable surfaces, namely, planes, cylinders, and cones, which form the basis of the families of projections (the projection families use the adjectival forms of the words: **planar**, **conic**, and **cylindrical**). Additional developable surfaces that comprise using multiple cones or cylinders are called polycylindrical or polyconic. The type of developable surface used in a mapped projection is commonly referred to as the **projection family** or **class.**

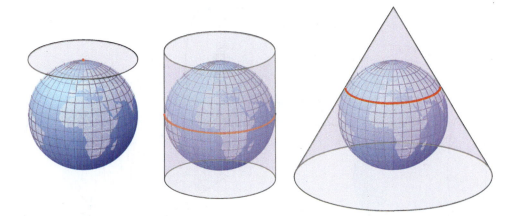

FIGURE 5.6
The developable surfaces (families) of projections: planar, cylindrical, and conic. These are shown in their normal aspects.

In the past, perhaps to save cartographers the painstaking task of developing new projections for each new mapping project, specific parts of the earth typically employed a specific projection class. The extreme latitudes, such as the polar regions, used a planar projection. High latitudes—between the tropics and the arctic circles—generally used conic projections, and areas near the equator used cylindrical projections. The traditional form of these was situated around the earth with a northern perspective, as shown in Figure 5.6. The angle at which the developable surface aligns with the earth is called the projection's **aspect**. These aspects, oriented along the earth's north-south axis, are referred to as **normal** aspects because these were the aspects most commonly applied in mapping. Projections rotated at precisely 90 degrees, such as a cylindrical projection wrapped around an east-west axis rather than the traditional north-south form, are called **transverse** projections (Figure 5.7).

Any other aspect falling between the two is termed an **oblique projection** (Figure 5.8). In the modern era, in which completely novel projections can be produced in a matter of seconds, the importance of the standardized or "normal" projection aspects for specific regions has been greatly diminished, but the legacy persists in both the terminology and the culture.

If you were building a projection and had decided which developable surface and aspect to use, the next decision you would make is about the light source. Recall Figure 5.5; from where would you base the hypothetical source of light which creates the projection? Once again, there is terminology to describe different "light source scenarios." The most common three are (1) **gnomonic**, with the light source at the center of the earth, (2) **stereographic**, from the point opposite side of the earth from the area being mapped, and (3) **orthographic**, from an infinite distance in the opposite direction of the

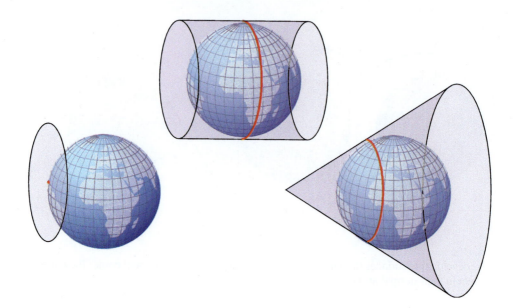

FIGURE 5.7
Transverse aspects of the three projection families.

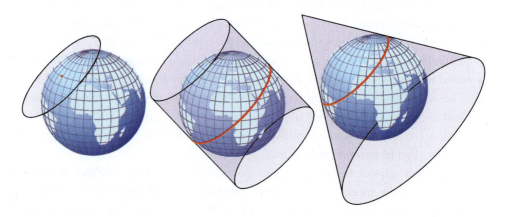

FIGURE 5.8
Oblique aspects of the three projection families.

point being mapped (Figure 5.9). Once again, each of these influences the types and patterns of distortion that ultimately appears in the projection.

Projections can severely distort the shape or area of a map. However, the points at which the earth's surface touches the developable surface *lack distortion of any kind*. This critical component of a map projection is called the **standard line** (or, in the case of some planar projections, technically it is a "standard point"). Distortion increases

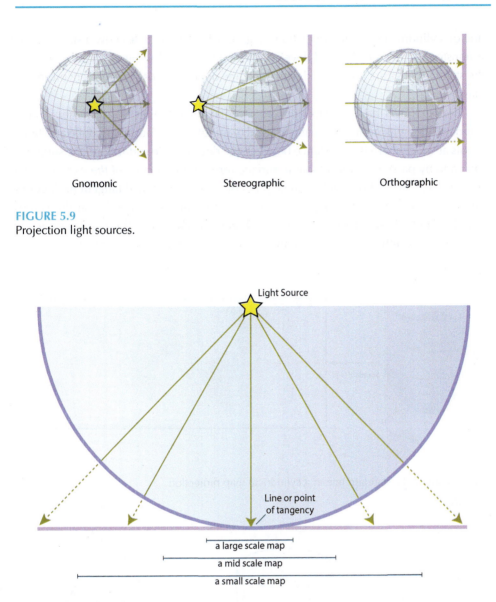

Gnomonic · Stereographic · Orthographic

FIGURE 5.9
Projection light sources.

Light Source

Line or point
of tangency

a large scale map

a mid scale map

a small scale map

FIGURE 5.10
An illustration of the effect of the standard line and distortion on a projected surface.

with distance from the standard line (or lines) on the map. If you think about this in terms of a literal projection, this makes some sense, as light must travel further with increased distance from the surface, and the distorting effect of the light is thus amplified (Figure 5.10).

Each of the projection classes tends to have a standard line with a distinctive shape. In the simplest cases, the standard line appears as a point in planar projections and as a line on cylindrical or conic projections. Figure 5.11 demonstrates how a standard line is transferred from a cylindrical developable surface. Examine Figures 5.6 through 5.8; the standard lines are highlighted with a red line or point across the different projection classes.

Cartographers have developed a way to maximize the lack of distortion around standard lines with additional mathematical manipulations. You can produce additional points at which there is no distance between the developable surface and the surface of the earth by passing the developable surface *through the surface of the earth's sphere* (Figure 5.12). Projections with a single point of contact are called **tangent projections**, while those which pass through the earth's surface or are embedded in the surface (yielding multiple standard lines) are called **secant projections**. This characteristic of a projection—whether tangent or secant—is called the **case** of the projection.

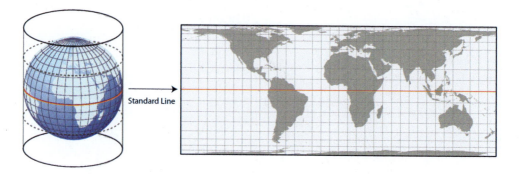

FIGURE 5.11
An illustration of a standard line in a cylindrical map projection.

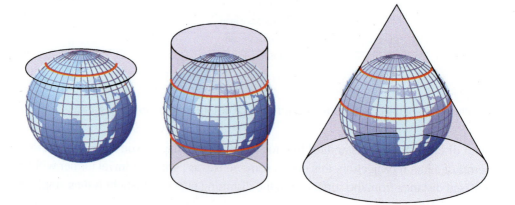

FIGURE 5.12
Examples of secant cases of the projection families. Standard lines are shown in red.

It's extremely useful to gain a solid familiarity with the projection families, types of developable surfaces, light sources, aspects, and projection cases. This terminology can be difficult to digest at first, but with a little experience, it does not take long for these concepts to feel intuitive. Many specific projections have this terminology built into their titles (such as the "Universal Transverse Mercator" projection or the "Gall Stereographic Projection").

The most critical component of choosing a good projection is having enough understanding of the ways that projections distort maps to guide your choices. The next section discusses the types of distortion that come with different projections and how you can pick the projection that works best for your map projects.

Major properties of projections

The two main types of distortion are called the **major properties** of a map projection: angle and area. If *area* is distorted, some parts of the world map appear much larger than they are in real life, compared to other parts of the surface. If *angle* is distorted, the shapes of countries and continents are altered. The two major properties are mutually exclusive; if a map completely preserves area, it is called an **equivalent** or **equal-area projection** and severely distorts shape. If the map projection goes the other way and preserves shape well, with distortions in area, it is called a **conformal** projection. Maps that fall somewhere in the middle and distort both area and shape to some degree are understandably called **compromise** projections (see Figure 5.13).

The minor properties of projections

In addition to the major properties, the **minor properties**, distance and direction, involve other forms of compromise. It is possible to preserve distance and direction from specific points on the map to all other points, but it is not possible to preserve these properties throughout the map.

Maps that preserve distance are called **equidistant** projections. In equidistant projections, distance can be preserved from one or two points to all other points along standard lines. A good example of an equidistant map is the equidistant conic projection (Figure 5.14). In this projection, the meridians have a constant scale, resulting in equally spaced parallels. Equidistant properties can only exist on compromise projections; it is not possible to achieve equidistance on conformal or equivalent projections.

Azimuthal projections are useful for specific navigational purposes, such as pilots' maps, and work well for mapping polar regions. Figure 5.15 is an example of an azimuthal equidistant projection; distance is true from the center of the map, the city of Moscow, to all other points. Connecting a line between Moscow and any other point yields a great circle route.

FIGURE 5.13
An example of an equivalent (Cylindrical Equal Area), and a conformal (Mercator) projection.

Projections in mapping

With so many options, you may wonder how one goes about choosing the best projection for a map. In the past, when maps were scribed by hand and projections were calculated manually, building a novel projection must have been a gargantuan and complex task. Thanks to modern advances in computing power and geographic software, the process of dealing with projections has become a lot easier. Cartographers are seldom stuck with a single projection, and it often now costs only a matter of seconds to try out and view different projections.

Large-scale maps

For large-scale maps that cover a small area, a carefully chosen projection with a standard line running through or near the mapped area should yield a map with practically no distortion—recall that there is no distortion at the standard lines and that distortion increases with distance from it. Every part of the map that covers a small area may be a matter of a few kilometers from the standard line because there is very little curvature across the earth in the map extent covering such a small area—have a look

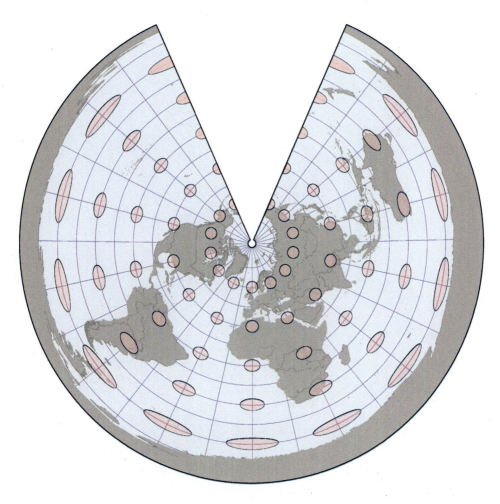

FIGURE 5.14
An example of an equidistant conic projection.

at Figure 5.10 once again; notice that how the portion of the earth covering a large-scale map is close to the shape of the developable surface to begin with. Put another way, the projection is represents a nearly flat section of the earth's surface, requiring relatively little distortion.

Intermediate scale maps

Even at intermediate scales—such as a map of a region or a small country—distortion can be minor and is unlikely to have a significant impact on the map. However, if you wish to produce an accurate map at that scale, you should choose a projection class and aspect that best fits the mapped region. The goal is to align as much of the mapped

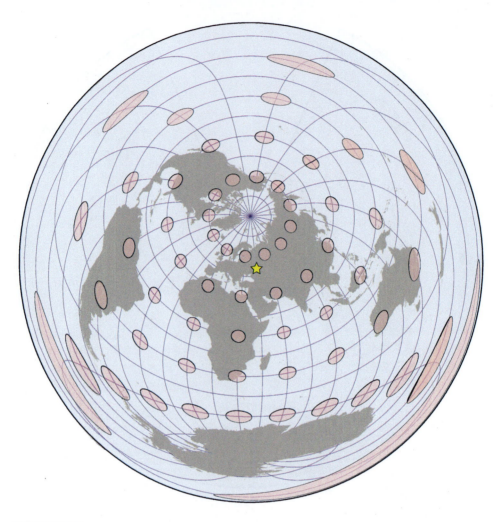

FIGURE 5.15

An example of an azimuthal equidistant projection based on the city of Moscow, represented on the map as a yellow star.

area as closely as possible to the standard line or lines. For a north-south-oriented country, such as Chile, it makes little sense to use a normal cylindrical or conic projection. A transverse cylindrical projection produces a north-south standard line that follows the length of the country (see Figure 5.16). The country is at an orientation tilted a bit eastward; it is not aligned perfectly with a north-south axis. In theory, you could refine the projection further by using an oblique projection to produce a straight line through the middle of the entire country to yield an even better fit, although the transverse projection should work well for most mapping applications.

Planar projections may offer a good solution for compact areas (such as France or Bolivia) because their standard lines are points (in the case of tangent projections)

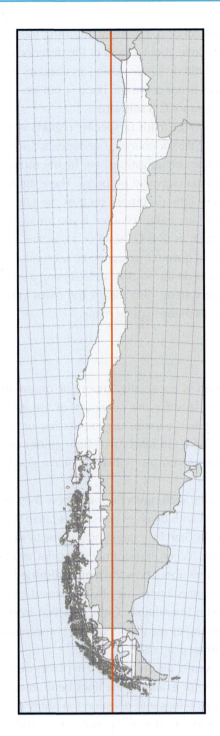

FIGURE 5.16
A transverse Mercator projection of the country of Chile.

or circles (for secant projections). Carefully adjusting the map projection so that the standard line forms a circle at about a third of the distance from the center of the map can produce maps with little distortion. Using an oblique, secant planar projection for areas outside of the poles is not common, but it is not too difficult to produce custom projections with contemporary GIS software.

Often the best solution for maps of specific regions is to follow the examples set by government agencies. State or county governments, such as the British or US Geological Surveys and military organizations, are highly motivated to produce accurate maps for navigation and management purposes. These agencies often use projections designed to produce very little distortion, such as the Universal Transverse Mercator (UTM) system, discussed later in this chapter. If you are tasked to produce a map with an intermediate scale, such as of a US state or a country in Europe, using whichever projection is used by local governance agencies is usually a good choice. In addition to relying on authorities to do the work of determining a suitably accurate projection, using official coordinate systems also means the maps you produce operate on the same coordinate system as many existing maps.

Small-scale maps

The choice of projection is especially important on small-scale maps that cover enough of the earth's surface for the curvature of the earth to produce significant distortion. In maps of the entire earth, large countries, or continents, you must make some tough decisions. There are lots of excellent guides online on selecting projections, and the GIS software you use is likely to have good descriptions in the help documentation.

As you become familiar with projections and how they distort area and shape, it is a good idea to experiment with different ways to visualize them. One useful technique is to examine the graticule. Start by examining a globe to see the true form of the lines of latitude and longitude. When transformed to a flat representation, the graticules can provide important clues about how the map is distorted across different regions of the world. The graticule intersections are always formed by lines going due east-west (parallels) and north-south (meridians). Graticules that intersect at angles other than 90 degrees indicate that angles are distorted, and you can safely assume that the map is not based on a conformal projection.

Some projections severely distort *direction* in places distant from the standard line. In a Robinson projection, for example, the lines bend toward the west or east as one approaches the poles. This brings into question the utility of including a compass rose or north arrow, since the orientation of "north" on the map depends on what part of the map you are viewing. It might make more sense to leave out the compass rose altogether in favor of including the graticule. The projection also distorts scale across the map, and so a single scale bar does a poor job of communicating the relation between map distance and real-world distance.

As noted previously, meridians are spaced at 111 kilometer apart at the equator but continually get narrower as they converge toward the poles. Parallels, on the other hand, are consistently at roughly 111 kilometer across the globe. Another sign of distortion of shape is when the parallels appear at variable distances or the meridians *are* consistently spaced.

The graticule leaves clues about other facets of the projection, such as to what class the projection belongs (Gede and Barancskuk 2015). For example, if the lines of longitude are straight, but the lines of latitude curve in an arc, it is probably a conic projection (Figure 5.17). If the lines of latitude are straight lines and perpendicular, it is a cylindrical projection. Planar (azimuthal) projections with a normal aspect are typified by circular lines of latitude. After you get some experience working with projections, the graticules will start to seem familiar and you should be able to recognize the basic projection class and aspect from the patterns on them.

A more immediately useful tool for examining distortion was introduced by a French cartographer named Nicolas Tissot in 1881 (Tissot 1881). The Tissot Indicatrix consists of several circles drawn on the map at regularly spaced intersections on the graticule. On the ground, each circle has the same shape and area. When the map is projected, the Tissot circles are distorted, depending on the properties of the projection. A map with Tissot circles that are of different sizes, but which maintain their circular shapes, has a conformal projection. Maps that show some of the circles as flattened ellipses, but which have the same area, use equivalent projections. Using GIS software, precise calculations can be performed on the ellipses to quantify the characteristics and extent of geographic distortion on the map (e.g., Goldberg and Richard Gott 2007). When you are working on small-scale maps with GIS software, it is helpful to keep a data layer of a Tissot Indicatrix on hand as a tool for examining distortion. Figures 5.13–5.15 include Tissot ellipses.

Choosing which map projection properties to emphasize

You should consider the map properties in tandem with the amount of distortion you need to accommodate as well as what it means for your reader. If the map is for navigation, showing property boundaries, or weather, emphasizing shape is important. While conformal projections preserve angle and shape to yield a familiar-looking map, they can severely distort area at the edges of the map, often near the poles in common conformal projections such as the Mercator.

If the sizes of areas on the map are important—such as any project that reports or analyzes aerial extent or maps that show population density with point symbols—then you should prioritize equivalence over conformality. Figure A.12 is an excerpt from a

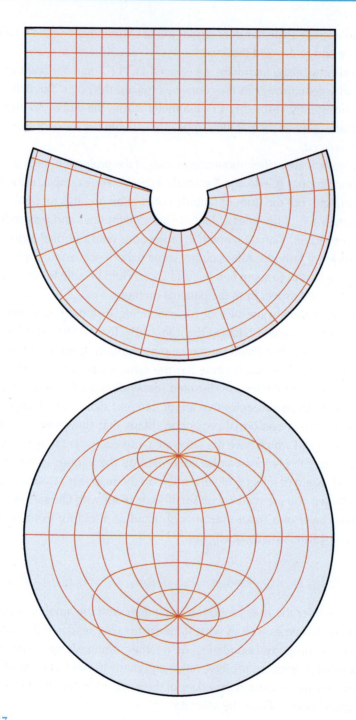

FIGURE 5.17
Examples (from top to bottom) of the graticule in a cylindrical, conic, and a planar projection.

paper reporting an analysis of changes in the coverage of conservation area around the globe. The map uses an equivalent Mollweide projection. Since the research was about measuring and comparing area, it was much more important that the maps provided an accurate representation of areas, and so an equivalent projection was the obvious choice. As is apparent on the map, the price of equivalence is that the shapes of the areas are severely distorted near the poles where features appear "scrunched up."

For a wide range of general thematic or reference maps, using a compromise projection that avoids extreme distortion of either of the major properties is often an ideal solution. Figures A.1 and A.8 are both thematic maps of the entire world and use a compromise projection.

Map projection examples

There are practically an infinite number of projections, ranging from traditional ones that have been used consistently for centuries, to the downright quirky. In this section, I will review several examples of global-scale projections in a discussion of their characteristics and potential uses. This selection contains projections developed by several key cartographers throughout history and represents some of the most common world map projections. As you read these, examine Figure 5.18., which shows basic examples of the projections, along with their graticules and Tissot indicatrices. Think about how the facets of the projection—the family, aspect, case, and light source—affect the appearance of the final map.

The Mercator Projection

The Mercator projection was popularized by Gerardus Mercator in the sixteenth century and may be the most famous among the global-scale projections. Mercator projections were the first to consistently appear in atlases and were often used in primary school geography classrooms throughout the world in the twentieth century. Mercator developed the projection for navigators, for whom it is well-suited because of its special property that any straight line is a **rhumb line**, a line that enables plotting a straight course with a constant bearing (a line of true direction).

The projection is cylindrical and normal, with the standard line that usually runs along the equator. Meridians are equally spaced straight lines, and parallels are gradually spaced further apart with distance from the equator. The distortions in area on this projection are massive. Areas in the high latitudes, distant from the standard line at the equator, are substantially enlarged. Greenland, for example, appears to be much larger than the entire continent of South America. The actual area of Greenland is about 2.16 million square kilometers, compared to South America's 17.8 million

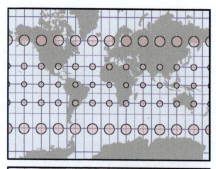

Mercator

cyllindrical, conformal

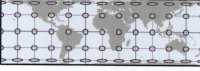

Lambert Cyllindrical Equal Area

cyllindrical, equivalent

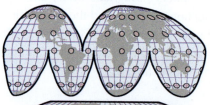

Albers Equal Area

conic, equivalent

Goode's Homosoline Interrupted

pseudocylindrical, equivalent

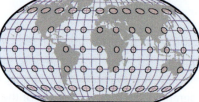

Robinson

polyconic, compromise

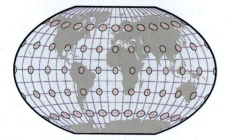

Winkel-Tripel

modified azimuthal, compromise

FIGURE 5.18

Examples of global-scale map projections with Tissot indicatrices.

square kilometers, about one-eighth the size—and yet the Mercator projection distorts area so severely that Greenland appears to be about twice as large as South America.

One critique of this projection is that because it exaggerates the size, it also exaggerates the importance of countries in the high latitudes, the generally wealthy countries in Europe and North America. It was often used during the Cold War to emphasize the importance and power of the Soviet Union, which dominates the map since so much of the land area was in the high latitudes. These concerns eventually led to schools (starting with schools in Europe during the 1980s) to adopt alternative projections that did a better job preserving area.

Cylindrical Equal Area Projection

When area is important to the mapping project, as is the case in many scientific mapping endeavors, preserving area is key. Some geographic software calculates area from the map, rather than the actual area it represents, and so working in equal area projections may be required to undertake a good aerial analysis. There are many different types of equivalent projections to choose from.

The Cylindrical Equal Area Projection was one among several projections developed and popularized by Swiss mathematician Johann Lambert in the second half of the eighteenth century (Lambert 1772). Distortions in shape are obvious near the poles. As a member of the cylindrical class of projections, the meridians appear as equally spaced, parallel lines. The "stretching" of the narrow areas between the meridians near the poles is compensated by compressing the distance between the parallels, which appear increasingly closer together towards the poles. The effect of this distortion on the map is that the high latitudes appear squashed. Due to the extreme way the map distorts areas near the poles, other equivalent projections with less obvious shape distortion are often preferable.

Albers Conic Equal Area Projection

A more visually appealing equivalent projection was proposed in the early nineteenth century by German cartographer Hans Christian Albers. The normal aspect of conic projections works well from the mid- to high-latitudes. The projection was commonly used in maps of Europe and was eventually embraced by mapping agencies in the United States to become a mainstay in the US government mapping efforts of North America. It is currently the primary projection used for maps of the United States in the USGS' *National Atlas* ([USGS] 2018). In order to minimize distortion, the USGS normally employs a secant case for the projection, yielding two standard lines, typically at 29.5 and 45.5°N (Snyder 1982, 95).

It is easy to tell that the projection is from the conic family because the lines of latitude are curved, invoking an image of the curved edge of a cone. The parallels are spaced the furthest apart near the standard lines, and the distance between them diminishes with distance from them. When the projection is used for a part of the world with appropriately selected standard parallels, the shape distortion is minimal, resulting in a map that has both accurate representation of area and a relatively appealing presentation.

Goode's Interrupted Homolosine

One option to minimize the amount of compression or shearing that occurs is to "tear" the map with an **interrupted projection**. John Paul Goode, from the University of Chicago, developed an interrupted projection that consists of several bands around the meridians that converge to the point. Because of the interruptions in the map, less areal distortion is necessary; the main drawback is that the map separates contiguous areas. The projection was popularized with the publication of *Goode's Atlas* in 1925 (Snyder 1987, 247, Goode 1925).

Robinson Projection

In the twentieth century, as world maps became more accurate and commonplace, cartographers continually sought compromise projections that were aesthetically appealing, but which did not distort area so heavily as to be seriously misleading. One of the most important and prolific cartographers of his time, Author Robinson, sought to develop a projection for displaying the entire world that did not draw attention to any extreme form of distortion (Morrison 2008). Although he originally named his projection an "orthophanic" projection (with the underlying meaning of "appearing or looking correct"), the projection is popularly called the Robinson projection. His projection was adopted for use in *Goode's World Atlas* and other publishers and organizations.

The map is pseudocylindrical with standard parallels at 38°N and 38°S latitude. A key advantage of this map is that little shape distortion occurs between the Arctic and Antarctic Circles. The curvature of the meridians toward the high latitudes reduces some of the area distortion apparent in traditional cylindrical projections, such as in the Mercator. The projection does not preserve any single property but produces a visually appealing map without extreme distortions of any map property. There are several other alternative compromise projections for world maps that have achieved popularity among difficult circles (for example, the Gall-Peters projection became popular in Europe, and the Winkel-Tripel projection was popularized by the National Geographic Society; both projections distort area less than Robinson), but the Robinson continues to be a favorite for general, global thematic mapping.

Winkel-Tripel

A similar compromise projection is the Winkel-Tripel projection. Developed years before the Robinson projection in the late nineteenth century by German cartographer Oswald Winkel, the projection was mathematically derived to simultaneously minimize distortions in area, direction, and distance (winkel.org 2020). One important distinction between this projection and the similarly appearing Robinson is that the parallels are not straight, and the shape of the earth is a bit more rotund, reducing some shape distortion. This projection was thrusted into prominence when the National Geographic Society adopted it as its standard projection for world maps in 1998.

The Universal Transverse Mercator coordinate system

The final projection discussed here is primarily used as a coordinate system rather than as a global projection for small-scale maps of the world. It is now used in a broad variety of government and official organizations for mapping applications that require a high degree of accuracy. The **UTM** system was developed by the US Army in 1947 for designating accurate Cartesian coordinates for the entire world (Snyder 1987, 57). As the title of the system implies, it is based on a transverse Mercator projection—a Mercator projection rotated 90 degrees so that the standard lines run north and south, rather than east and west. The zones are numbered from 1 to 60, starting at the International Date Line and then continuing eastward (Figure 5.19).

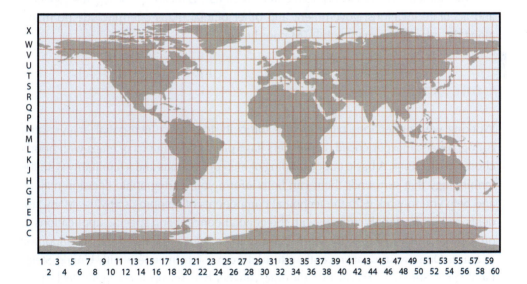

FIGURE 5.19
A map that shows the UTM coordinate system zones.

To achieve accuracy, the system divides the world into zones that are six degrees wide each, with a transverse Mercator projection oriented so that standard lines run through the middle of each zone. Each zone is then given a unique coordinate system, designated in meters. The designers of the system wished to avoid negative numbers, and so they included **false eastings** and **false northings** as the origins of the system. In the northern hemisphere, the equator is designated as the "x-axis," representing the line from which the distance to the north is measured, and the "y-axis," the line from which points to the east are designated, and is 500,000 meters (50 kilometers) to the west of the central meridian of each zone. In the southern hemisphere, the equator is designated as 10,000,000 meters (10,000 kilometers) to achieve the same effect of avoiding negative numbers. Zones are also divided into north-west bands that are eight degrees of latitude, each using letters to designate them, starting with "C" at 80°S and progressing northward. The letters "I" and "O" are skipped to avoid confusion with the numbers one and zero, respectively. To achieve the highest levels of accuracy, different zones in the UTM system are based on reference ellipsoids that most closely match the geoid for that zone. The polar regions use a different system called the **Universal Polar Stereographic (UPS)** system that uses an azimuthal stereographic projection to build a similarly accurate coordinate system.

While the transverse Mercator cannot be used to produce highly accurate maps for large areas, the coordinate system is widely employed by official and government institutions around the world. If you are producing a large-scale map that fits within a UTM zone, adopting the coordinate system and projection is a straightforward and effective solution for your maps. To facilitate their use as means for designating specific and accurate locations, government and other official maps often include tic marks along the edge of the map with UTM coordinates.

Conclusion

Working with projections often creates difficulty for novice geospatial analysts and cartographers; inexperienced mapmakers have produced millions of maps with poorly chosen projections. A basic understanding of map projections and a familiarity with the terminology that surrounds them can both inform your cartographic decision-making and help you to work with the technical components of mapping software. Many students and professionals struggle with the general concept and specific applications of projections in cartography for years before they master the ideas, but it is important to learn. One my students recently returned from a job interview with United States Geological Survey to report that the interviewers had set up a computer to test his ability to manipulate map projections! Fortunately, modern computing makes working with projections much easier and straightforward than it was in the past.

Discussion questions

1. Think about describing location. How do you describe the location of yourself or your house? What are the "anchors" that you use (to what features around you do you reference your location)?
2. Pick a topic and a country on Earth that you might be interested in mapping and think about some of the general purposes of your map. With the help of the Internet or other sources, pick a projection to use in the project and justify your choice.
3. Examine the maps in the map gallery included in Appendix 1. Look at the maps with an eye for the types of projections the maps used. Are you able to identify specific projections? Would you suggest altering the projection on any of the maps?
4. Examine the projections displayed in Figure 5.18. Which of the projections seems the most natural or otherwise appropriate for general mapping purposes? Discuss your choice and your reasons for it with your peers (or with someone who does not have any cartographic experience at all). What were your others' rationales for their choices?

References

Gede, M., and A. Barancskuk. 2015. "Determining the Projection of Small Scale Maps Based on Graticule Line Shapes." *10th Jubilee Conference + Workshop Digital Approaches to Cartographic Heritage*, Corfu, Greece, May 27–29, 2015.

Goldberg, D.M., and I.J. Richard Gott. 2007. "Flexion and Skewness in Map Projections of the Earth." *Cartographica: The International Journal for Geographic Information and Geovisualization* 42 (4):297–318. doi: 10.3138/carto.42.4.297.

Goode, J.P. 1925. *Goode's School Atlas; Physical, Political, and Economic: For American Schools and Colleges, by J. Paul Goode*. Chicago, New York: Rand McNally & Co.

Lambert, J.H. 1772. *Beiträge zum Gebrauche der Mathematik und deren Anwendung*. Berlin: Verlage des Buchladens der Königl. Realschule.

Morrison, J. 2008. "Arthur Howard Robinson, 1915–2004." *Annals of the Association of American Geographers* 98 (1):232–238.

[NASA], National Aeronatic Space Administration. 2017. "Earth Fact Sheet." Last Modified March 16, 2017, Accessed August 25, 2018. https://nssdc.gsfc.nasa.gov/planetary/factsheet/earthfact.html.

Snyder, J.P. 1982. "Map Projections Used by the U.S. Geological Survey." In *Geographic Survey Bulletin 1532*, 313. Washington, DC: US Department of the Interior.

Snyder, J.P. 1987. "Map Projections: A Working Manual." In *Professional Paper*, 57–247. Washington, DC: US Geological Survey.

Tissot, N.A. 1881. *Mémoire sur la Représentation des Surfaces et les Projections des Cartes Géographiques*. Paris: Gauthier-Villars.

[USGS], United States Geological Survey. 2018. "The National Map Small-Scale Collection." Accessed July 6, 2018. https://nationalmap.gov/small_scale/.

[USGS] United States Geological Survey. 2020. "What Is a Geoid? Why Do We Use It and Where Does Its Shape Come from?" Accessed January 27, 2020. https://www.usgs.gov/faqs/what-a-geoid-why-do-we-use-it-and-where-does-its-shape-come?qt-news_science_products=0#qt-news_science_products.

winkel.org. 2020. "Winkel Tripel Projections." Accessed June 27, 2020. http://www.winkel.org/other/Winkel%20Tripel%20Projections.htm.

6 | Text and typography

While the study of cartography tends to focus on the visual elements of design, most successful maps make liberal use of text. Many important map elements convey critical information with text, including the title, subtitle, authorship, legend, and, of course, map labels. In many instances, narrative prose to describe the background or context of the map can add meaning to the map itself.

The use of text in cartography can be divided into basic categories: (1) contextual and background information about the map, (2) information to help map readers understand the meaning of the map and interpret the data presented on it, and (3) spatial information itself, such as the names and locations of features, presented in the form of **map labels**. The use of text in your maps introduces an additional set of cartographic decisions, some of which can have important effects on how well the map achieves its goal. Text can make an otherwise mediocre map into an effective and compelling one. This chapter begins by introducing the concepts and terminology around text and concludes with a discussion of good practices for using text in your maps.

The characteristics of text

Typography, the study of text and its appearance, began with the first instance of formal written language over 5,000 years ago in ancient Phoenicia. Over the years, through the Middle Ages, when formal writing and calligraphy developed, following the invention of the printing press and into the modern era when text is mostly produced using computing technology, the styles and form of text have continually evolved. The goal of typography is to communicate written information in a clear, effective, and visually appealing manner, not unlike the goals of cartography itself. Knowing about the basic tools and options for manipulating the form and style of text can help you enhance the creative potential of your map. Familiarity with the concepts of typography will help you think through cartographic decisions as well as to facilitate using modern graphical and layout software.

Text size

One of the most important facets of text for cartographers is its size. If you have used a word processer, you may have noticed that text size is measured in **points**, equivalent to 1/72nd of an inch. The text size refers to the distance between the lowest points of any of the letters (such as the loop of a "g" or the descender on the "j") and the highest point (such as the top of capital letters or the ascender on a "b"). Text intended for blocks of narrative or prose, such as most of the text in this book, is usually around 10 or 12 points.

The size of the text can serve as a critical indicator of the importance of map elements or features, and you can use the text size to direct readers' attention. It is difficult to provide consistent guidelines for what point sizes to use because the appropriate size depends on the font, line thickness, and style of the text. The elements that you want the readers to immediately notice with their initial scan of the map, perhaps the title or subtitle, should be featured with a prominent text size, perhaps around 30 or 40 points for a standard page-size layout. Consequently, choosing the size of the text in your map requires some judgment.

Less important elements, that is, information that is necessary to include on the map but is less important for the map reader to notice immediately, should generally be as small as possible while remaining legible. It is usually a good idea to include the source of the data on the map, for example, but you also do not necessarily want to draw attention to it. Reserve the reader's attention for the story you want to tell.

The text size can provide a clear path through the visual hierarchy on your map. As an example, look at Figure A.2 in the map gallery, a map of place names in Minnesota. The first thing that the eyes are drawn to is the title, "French vs. All Place-Names in Minnesota." This headline gives you a basic overview of the main idea behind the map. The next objects you are likely to examine are the map itself and the legend. The map legend has the second-largest text on the map and is easy to find, along with the labels for Minneapolis-St. Paul, the major cities in the state. Other text provides additional information and context for the map, such as the labels of the surrounding states and some key lakes.

Font (typeface)

The **typeface** (also often referred to as the "**font**"; both terms are used here interchangeably) of the text refers to the style and shape of the letters. The history of typefaces is deep, extending back to the very origins of writing. Over time, as writing systems developed and evolved, different fonts moved in and out of style, some of which continue to have cultural and design relevance in the modern era (Sacks 2003).

The two most commonly used font families are **serif** and **sans serif** ("sans serif" is French for "without serif"). **Serifs** are small finishing strokes on the letters to give them a bit of artistic flair. The first known example of a serif font was found on an ancient Greek engraving dating from 334 BC, after which the practice was picked up by the Romans who used serifs frequently in stone inscriptions (Sacks 2003, 107–108). It is not clear why serifs were originally introduced, but one idea is that the first stone carvers mimicked brush patterns as they inscribed the text on stone (Catich and Gilroy 1991). Although it was most certainly not the intention of the original designers, a convenient advantage of serifs is that they provide the eye and mind something to anchor upon. When you read text, your mind is performing an incredible number of complex functions, one of which is to keep track of where you are viewing (particularly when you scan from one line of text to another). Serifs can therefore serve to reduce the cognitive load in the readers' minds. Some common examples of serif fonts include *Times New Roman*, *Garamond*, *Palatino*, and *Century* (Figure 6.1).

Sans serif fonts, as the name implies, lack serifs and generally appear more modern. They were once widely applied to computer applications because computer screens were not able to display the detail of serif lettering effectively. This was due to the comparatively low resolution of computer screens and difficulty with **anti-aliasing**, the practice of blending colors on computer screen so that the text appears to blend in with a continuous background, rather than showing "pixilation" or blockiness (Figure 6.2). Consequently, the traditional advice is to stick with sans serif fonts for text intended for reading on a computer screen. Advances in computer and display technology have largely rendered

Times New Roman

Garmond

Palatino

Century

Poor Richard

FIGURE 6.1
Examples of serif fonts.

FIGURE 6.2
A demonstration of anti-aliasing and halo effect.

this advice irrelevant because modern monitors with high resolutions are able to display just about any font quite well. Some san serif fonts are elegant and easy to read, making them good choices for mapping work. Common examples of sans serif fonts include *Arial* (very similar to *Helvetica*), *Calibri, Myriad, Verdana,* and *Corbel* (Figure 6.3).

Decorative fonts are a large group of special fonts with "strong" personalities, often designed to convey a specific theme or to evoke an emotion. Decorative fonts are generally not suitable for blocks of text—they tend to be complicated and difficult to read—and are best reserved for signs, headlines, and titles. If you pay attention to fonts, you might observe a wide range of decorative fonts used in advertising, appropriately selected to sell anything ranging from herbal remedies, beer, or military clothing, to sophistication. Fonts in this family may be either serif or sans serif.

Examples of decorative fonts include *Charlemagne, Stencil,* and *Old English* (Figure 6.4). Each of these arguably evokes a place, time, or concept, which is often implied by the font name itself. *Charlemagne,* named after the Frankish king from the seventh century, evokes the idea of the Middle Ages because this style was common in the era. *Stencil,* perhaps due to the widespread use of stencil lettering by military groups, compels one to think about the military. *Rosewood* echoes lettering from the "Wild West" of the nineteenth century USA. In serious mapping, you should generally avoid using decorative fonts, but when applied to the right purpose and audience, it can effectively evoke an idea or concept.

Related to decorative fonts is a class of fonts called **script fonts**, designed to mimic written scripts (Figure 6.5). Some designers and font dealers (there are indeed

Arial

Calibri

Myriad Pro

Verdana

Corbel

FIGURE 6.3
Examples of sans serif fonts.

CHARLEMAGNE

ROSEWOOD

Old English

STENCIL

Papyrus

FIGURE 6.4
Examples of decorative fonts.

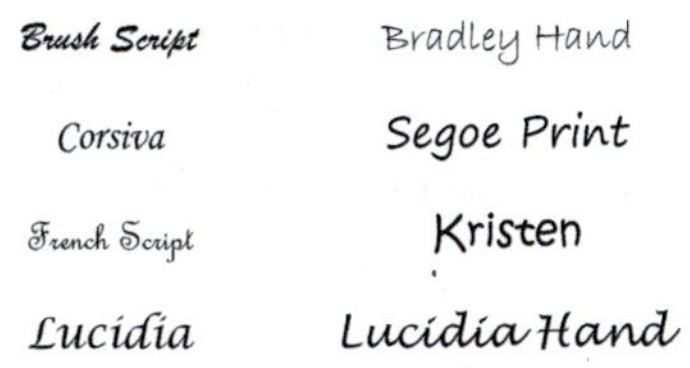

Brush Script   Bradley Hand

Corsiva   Segoe Print

French Script   Kristen

Lucidia   *Lucidia Hand*

FIGURE 6.5
Several examples of script fonts are shown here. The column on the left shows "formal" script fonts and the column on the right shows "informal" or "handwriting" fonts.

organizations that design and sell fonts) classify script fonts into "formal" and "informal" subclasses. The left column in Figure 6.5 shows formal script fonts and the column on the right shows informal or "handwriting" fonts. You might recall seeing formal-looking script fonts in wedding invitations or other places where the author wishes to imply sophistication, such as in titles of romance novels. Less formal-looking script fonts can be spotted on signs for restaurants, cafes, or specialty stores. Some script fonts are notoriously difficult to read beyond a few words at a time, and they often must be enlarged substantially to be legible. As with decorative fonts, script fonts should generally be avoided in mapping unless there is a compelling reason to use them and they can be printed in a clear and legible form.

A final type of fonts is called **monospaced**. Monospaced fonts were designed for computer applications and are so named because the distance between each letter is consistent (Figure 6.6). In monospaced fonts, for example, the letter combinations "ij" take up the same amount of space as "HX." This type of font is particularly useful for programs that use text to convey statistical output, such as a simple histogram or box plot. Some of the first digital maps used monospaced fonts to build entire maps, built from special characters, such as "*,' "∧,' and "/" (Figure 4.5 is an example of a map built from a monospaced font called "courier"). Unless you are working with software that requires the use of such a font, there is little reason to prefer them over fonts from the other classes on modern maps. Monospaced fonts are defined entirely by their spacing and may belong to any of the other classes.

Like much in cartography, the guidelines for font selection in mapping generally follow the basic rules of common sense. In most cases, you do not want your reader to notice the font (you would rather they notice the map and its content), but rather it should serve as a tool to effectively convey important information. Be careful to examine all fonts for legibility; some fonts include small line thickness for parts of the letters, making them difficult to read in small to moderate sizes.

It is good practice to use the same font throughout the map and all its elements, though some good maps can include multiple types of fonts to distinguish between different parts of the map. Nearly all the example maps from the gallery in this book use a single font. One exception is the map of Europe in 1911 (Figure A.4) which uses

Verdana

Courier

Lucidia Console

Consolas

FIGURE 6.6
Examples of monospaced fonts.

a sans serif font for the title, page numbers, and labels of mountain ranges, and a serif font for the other labels and elements on the map.

Some publishers have strict guidelines that impose the use of a specific font, but you may have the option to be creative. A good practice in cartography and design work is to build a font sheet, similar to Figures 6.1, and 6.3 through 6.6 (showing lists of fonts) for you to refer to as you design your map. If you want the map to appear informal, for example, chose a font that conveys the idea. If you are unsure about what a font communicates or its effect on the map (such as its impact on legibility), never forget you can always ask a peer for his or her opinion.

Type style

Type style modifies fonts by increasing the line thickness (**bold** or **boldface**) or by sloping them to the right (**italics**) (Figure 6.7). Strictly speaking, type styles are different fonts, but they are nearly always grouped together under a single font name in software applications. Italic fonts are frequently used to emphasize special words, such as titles, quotations, or words borrowed from other languages. Bold font makes the line width of the letters thicker, often used to emphasize words or phrases (such as the freshly introduced words and concepts in the prose of this book).

There are some conventions in mapping about when to use italics. It is common in reference maps, for example, for the labels of bodies of water to appear as italics, a practice adopted both by the USGS and the British Ordnance Survey in their topographic map series. Several examples from the map gallery (Appendix 1) make use of this convention. Bold text can be used in maps to emphasize important elements or subsets of features, similar to how you might use text size. In a map that shows cities, for example, the names of large cities or capitals can be printed in bold. Do not hesitate to use bold text for the entire map if it makes it easier to read and does not give the text more visual weight than desired.

Text spacing and orientation

Because labels are an important part of many maps, text gains an additional function beyond merely conveying words to the reader; they can communicate *spatial* information

Regular (Arial)	Regular (Times)
Bold (Arial)	**Bold (Times)**
Italicized (Arial)	*Italicized (Times)*
Bold italicized (Arial)	***Bold italicized (Times)***

FIGURE 6.7
Examples of different type styles using common serif and sans serif fonts.

about the locations of things. Consequently, it is important to understand the ways you can manipulate spacing and orientation in text to communicate effectively.

Kerning and tracking

Kerning is the "adjustment of horizontal space between individual characters in a line of text" (Adobe 2018). When publishing was the domain of the printing press, the angles on the blocks of individual letters had to be adjusted so that they could overlap neighboring letters appropriately. For instance, if an "A" was set next to a "V," the type blocks had to cut at an angle so that they could overlap (Figure 6.8); if a square block was used, the letters would appear inconsistently spaced or too far apart. Kerning referred to the specific adjustments to the edges of particular pairs of letters. The development of word processing and typesetting software has made the process much easier, and now most software perform some degree of kerning automatically. Some sophisticated applications enable users to adjust the kerning values between individual characters.

A similar and related term is **tracking**, or "letter spacing" the distance between the characters in text (Figure 6.9). Compare the name of the Welsh village, "Llanfairpwllgwyngyll," without any tracking to the same name with some additional tracking added

AV AV

FIGURE 6.8
An illustration of kearning.

(em = -50)	Llanfairpwllgwyngyll
(em = 0)	Llanfairpwllgwyngyll
(em = +100)	Llanfairpwllgwyngyll

FIGURE 6.9
Examples of tracking with indicated em values.

("L l a n f a i r p w l l g w y n g y l l"). Most word processor and graphics applications enable tracking. Units of tracking and kerning are nearly always specified as a portion of the distance of the current text size. *Adobe* products, for example, uses the unit "em," which is 1/1,000 of the current point size of the text. If you are using 20-point text, for instance, specifying tracking of 100 will add an additional two points (since 100 is one-tenth of 1,000, and the point size is 20) between the text.

Manipulating tracking can help to make text legible on the map, particularly for text with low point sizes. In some cases, best applied with very large text, it can be helpful to reduce tracking to make the text fit or improve its balance.

Line spacing

In addition to adjusting the spacing between individual letters and words, you can adjust the **line spacing**, the vertical distance between the lines in a block of text (Figure 6.10). Line spacing is measured in points, from the baseline of each line to the lines above and below it, and can be specified either as a point value or as a percentage of the point size of the text. A value of 100% in 12-point text means that the lines above and below start at exactly 12 points above or below the baseline of the text. This is too tight for most applications since it leaves no room between the lines. Most

100% line spacing

A map is a drawing or other representation of the earth's surface or a part of it made on a flat surface, showing the distribution of physical or geographical features (and often also including socio-economic, political, agricultural, meteorological, etc., information), with each point in the representation corresponding to an actual geographical position according to a fixed scale or projection; a similar representation of the positions of stars in the sky, the surface of a planet, or the like.

120% line spacing

A map is a drawing or other representation of the earth's surface or a part of it made on a flat surface, showing the distribution of physical or geographical features (and often also including socio-economic, political, agricultural, meteorological, etc., information), with each point in the representation corresponding to an actual geographical position according to a fixed scale or projection; a similar representation of the positions of stars in the sky, the surface of a planet, or the like.

150% line spacing

A map is a drawing or other representation of the earth's surface or a part of it made on a flat surface, showing the distribution of physical or geographical features (and often also including socio-economic, political, agricultural, meteorological, etc., information), with each point in the representation corresponding to an actual geographical position according to a fixed scale or projection; a similar representation of the positions of stars in the sky, the surface of a planet, or the like.

FIGURE 6.10
Examples of different line spacings of the Oxford English definition of a map. The text in this example is ten points.

typographers recommend that text should be somewhere between 120 and 150%. Increasing the line space gives the text a bit of breathing room, but you will generally want to strike a balance between legibility and space to make the text appear balanced and easily legible. If you are working with a block of text on a map, you can adjust the line spacing a little bit to improve the balance of the map, but you should ensure that it does not distract the reader.

Orientation

Changing the orientation can be an extremely helpful tool for labeling features to accommodate spaces on the map, convey information about the names of areas on the map, or to clearly associate the text with a feature on a map that does not appear horizontal, such as a river.

With a few exceptions (mostly from Middle Eastern languages, such as Hebrew and Arabic), text is presented from left to right, and from top to bottom. When you are reorienting text to a non-horizontal direction, it should first maintain a left-to-right orientation and, following that, a top-to-bottom orientation. Never orient text in such a way that requires your reader to read text backward, from right to left, or from bottom to top. If you have a situation in which you wish to orient text along a feature that is oriented to the top and right (such as a river in a northeast-southwest direction on a map), prioritize the rule of requiring your reader to go from left to right over the top-to-bottom rule.

Special effects

Anyone who has spent much time browsing the Internet (or paying attention to advertising on billboards, shop signs, or flyers) can appreciate the variety of ways in which text can be altered with special effects. Just about any effect that can be applied to a graphic can also be applied to text. Some of these effects can be useful in cartography, but, as ever, it is important to avoid using them in such a way that distracts the reader or interferes with the legibility or purpose of the map. Novice cartographers may be tempted to include special effects to make the map interesting, but these effects should generally be used in a subtle manner to avoid overwhelming the map with unnecessary graphical noise.

Halos and masking

A **text halo** is simply an outline around the text itself that is a different color from the text (see Figure 6.11). Text halos are usually specified in the same units of points used to indicate text size. The technique can serve multiple purposes in mapping: to highlight

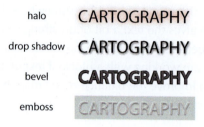

FIGURE 6.11
Examples of special effects on text.

important text, to provide contrast with the layout space or map itself, to cover up the features underneath the text so that the text does not interfere with the map features, or to make text generally more legible. An alternative usage of the term "halo," as noted earlier, refers to the effect that occurs when a letter is designed for anti-aliasing on a background with a specific color, but is then moved to a background with a different color, seldom a problem in modern map work.

In some cases, adding a halo to text in a map can make the feature stand out, nudging the text a bit higher in the map's visual hierarchy. It is generally only appropriate to use halos in this way for the main title of the map or perhaps a few key labels; there is seldom any reason to emphasize the other textual map elements (such as the legend heading, data source, or authorship) in this way.

Drop shadows

A drop shadow emphasizes the text by adding a copy of the letter offset by a few points, usually with a lighter or semi-transparent color than the letter itself (see Figure 6.11). If the graphics or mapping program you are working with is raster-based, you may be able to apply a variety of special effects to the shadow, such as blurring and transparency. Drop shadows are generally effective ways to highlight text and can produce elegant-looking titles and signs if they are appropriately specified. I recommend adding drop shadows without much distance from the text; it is usually a good idea to keep the shadows within 10% or 20% of the text size. Giving the drop shadows too much distance can yield the "double-letter effect," which makes it appear as if two letters are printed. In mapping, shadows can serve a similar function to halos and can help with improving the contrast and legibility of the maps.

Other graphical effects

Most graphical software enable a variety of other graphical effects, but few of these are particularly relevant or useful for maps. Examples of such effects include **beveling**

and embossing, giving text the appearance that it is slightly extruded. Color overlays, pattern fills, and other variations in the aforementioned techniques can enable an impressive number of creative ways to present text. If you have the opportunity to become familiar with graphical software, it is worth experimenting with some of these tools, but always bear in mind the goals of including text and remember that clear and crisp designs with few graphical bells and whistles are usually the most effective methods for mapping. It's worth mentioning that none of the text in the maps in the map gallery included in the book (Appendix A) use these kinds of graphical effects.

Guidelines for labeling maps

Map labels serve to convey information about the names of features on the map itself, often serving as "multifunctioning graphical elements," by communicating not only the name of the feature they represent but also information about the location of the entity being mapped. While labeling maps previously involved careful and often painstaking manual typesetting, advances in modern digital cartography have alleviated this task considerably through automation. Modern GIS software has built-in tools to easily place labels throughout a map. Depending on the map project and how much time you want put into the map, you may choose to manually add and adjust labels yourself.

Although there are many books and articles that describe good practices in labeling map features (e.g., Brewer 2005, Peterson 2014), there are few universally accepted principles. Using a bit of basic common sense and keeping to the general design principles of cartography, however, there are several guidelines you can follow.

Include labels only when it serves the map purpose

Before you choose whether to include map labels in the first place, first consider whether it contributes to the goal of the map. In the map of place names in Minnesota (Figure A.2 in the map gallery), for example, the author has included labels only sparingly. She labeled Canada, the surrounding states, Lake Superior, and the primary city of the state to orient the reader to the general location and context of the map. Although the very topic of the map is about place names, none of the other place names are included. The point of the map is to show different types of place names and compel the reader to think about the patterns in them, not to show the reader what those places are actually called, which would make the map far too cluttered. Figure A.3, the map of solar power potential in the state of Missouri, uses an analogous technique by including only key labels to orient the reader to some features (roads and towns) that may be familiar to them. Other examples from the map gallery (e.g., Figures A.8 and A.12) do not include labels at all, as the reader is assumed to have enough familiarity with the geography, so that explicit labels are not needed. Labeling specific geographic features does not serve the purpose of many of these thematic maps—to communicate spatial patterns in data (such as cases of tuberculosis and biological conservation).

Reference maps (such as Figure A.4), whose purpose is to provide users with a general overview of an area, are a different story. Labels are practically an inherent component to reference maps, and so the key decision is which features to label and how to label them. At the end of the day, a balance must be struck between providing good information and providing useful and consistent representation of the feature names, while ensuring that the labels can be easily read.

Make the labels legible

If the labels are difficult or impossible to read, the very point of including them on the map is defeated. A jumble of characters the map user is not able to read reduces text to mere graphical noise. While it often takes a matter of only seconds to add thousands of labels to a map using modern software (and it might seem tempting to get so much done so quickly!), the risk of using those tools is that the map easily become overloaded with too much unnecessary information.

You can review a short checklist to ensure that your text is legible to a wide range of readers: (1) the text should be large enough to read, (2) the text should have enough contrast with its background to read, (3) the orientation and spacing of the text should be designed to make the text as easily legible as possible, (4) labeling should be systematic and consistent, and (5) when feasible, there should be sufficient negative space around the text to reduce the amount of cognitive work the reader has to take on to comprehend them.

While it may seem self-evident that text should be large enough to read, there are plenty of examples of maps with text that is too small. The smallest legible size depends on several factors, such as the font, style, spacing, placement of the text, and the ability of individual map readers to read fine print. While you can always judge yourself whether a label is legible, it is often helpful to show the map to the people who make up your intended audience: ask them to read you some of the features on the map.

A common scenario leading to text that is too small is when the map ends up being displayed or viewed at a different size than the original designer intended. If you produce a poster-sized map layout for a conference presentation, the labels need to be large enough to read from a distance, which means that the smallest labels on the map could be as large as 40 points. If that poster then ends up winning a cartography competition, it will inevitably show up on a web page, where it will be viewed on a single computer screen, where the 40-point text could be rescaled to a size far too small to read. It is often incumbent upon the cartographer to imagine and plan for the different sizes and formats in which a map will ultimately appear.

A second guideline to bear in mind for legibility is contrast. While it is often possible to read low-contrast text, such as black text on a somewhat less-dark background, low-contrast text means that the reader must spend extra cognitive power to read it. Most of the text that we read, such as in this book, is shown with the most contrast possible—dark or near black letters on a white or near-white background. Producing sufficient contrast in map labels can be a challenge because they must coexist with a rich variety of other colors and graphical symbols. Traditional black labels become difficult to read on dark backgrounds. Take a second to examine the labels on the map of solar power resources in Missouri (Figure A.3), for example. The southwestern part of the state is shaded with a dark red color to represent the relatively high solar power potential in that part of the mapped region. The map labels for the cities of Joplin and Springfield are much more difficult to discern than cities in the northeastern part of the state, such as Kirksville, colored with a lighter shade of orange to represent lower solar power potential.

There are several techniques you can use to generate sufficient contrast. For maps in which there are few labels, where the labels are extremely important, or where there is not a lot going on with the data, adding text boxes—a graphical area behind the text that blocks out the underlying symbology—can produce an acceptable solution.

The general practice of blocking out features on a graphic is called **masking**. Cartographic lines that interfere with the lines of the label's characters can make text on the map extremely difficult to read. Take the example of text that is meant to label the name of a municipality on a map with lots of roads (Figure 6.12). The background of the map is light gray, and both the text and the primary roads are black. By applying a simple four-point halo around the text with a color slightly lighter than the background in the

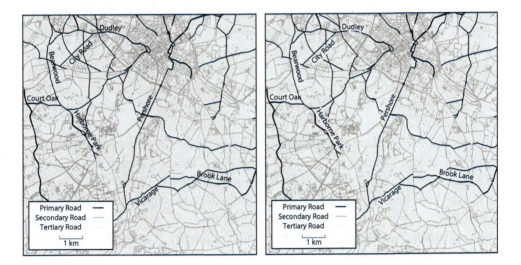

FIGURE 6.12
Examples of masking for labels in a map of roads in Birmingham, UK.

125

map on the right, the lines of the underlying roads are essentially blocked out, giving the labels some breathing space and avoiding interference from the features. Another type of masking could include blocking out the underlying graphical features with small graphical rectangles behind the text, filled in with the same color as the background.

Producing contrast with consistent labeling can be particularly difficult when the map contains a variety of contrasting colors. Halos can be used to ensure that some part of the outlines of the letters has strong contrast with the background. By including a light-colored halo around dark-colored text, for instance, all parts of the text will have enough contrast—either through the halo or the text itself—with some part of the background.

The third component of legible text is to make the text spacing and orientation easy to read. As noted earlier in the chapter, if you reorient the text to a different angle or a curved line, always ensure that it can be read from left to right, and, following that, when possible, from top to bottom. If the text appears as too tight or jumbled along a curved line, increase the tracking to ensure that there is enough space for each letter to be easily viewed.

Manipulating the size, contrast, and style of the text can serve as a convenient and graphically succinct way to communicate spatial information. The visual hierarchy of labeled features can be codified through your decisions about typography. The most important features at the top of the hierarchy should be the easiest to read, such as the country labels in the reference map of Europe (Figure A.4). If the label is important enough to include it on the map, it should certainly be possible to read, but it is okay to make the reader work a little harder to read the less important features.

It is critical that the features belonging to a common class are labeled consistently in order to leverage the natural human ability to discern patterns. Even without any explicit reference to the meaning of map labels, a lot of information can be implicitly communicated if the styles are consistent. If all the labels of major bodies of water are printed in 12-point italicized text with all capital letters, readers who want to explore the names of large bodies of water will quickly learn to seek out other text with a similar style. If you label rivers, towns, or any other map feature in a specific way, ensure that you use a consistent style for all the features of that class.

A common problem in map labeling is avoiding interference between the labels and the graphical symbols on the map (often the very feature the label should refer to). If road, boundary, or railroad symbols interfere with the label of a river, for example, the river label becomes extremely difficult to read. While this sort of interference is often difficult or impossible to avoid altogether, there are some techniques you can employ to mitigate the problem. Under no circumstances is it acceptable for the labels to interfere with other text.

The best option is usually to move the label around so that it does not interfere with any of the map symbols. While there are some conventions for label placement, it is usually better practice to give the label a less-than-ideal position rather than allow features to interfere with it. One option is to selectively omit the label of the interfering feature. The automated labeling tool in *ArcGIS*, for instance, enables users to specify which features have the greatest priority; users can input a list of features and labels in order of importance. If the software detects that there is a conflict (that the objects overlap or are too close together), it will keep the highest priority object and remove the ones lower on the list. Often a more satisfactory option is to use masking to block out the underlying features to generate space around the text as described above.

Communicating the target feature

When you are giving names to objects in maps, it is important that there is a clear and unambiguous connection between the label and the feature to which it is intended to refer, known as the **target feature**. Scholars in cartography have compiled rules for the placement of labels (Imhof 1962), and, more recently, have begun to develop algorithms for automated placement of labels in digital cartography (e.g., Freeman 2005, 2007).

The ideal placement for a label on point features is close to the feature, above and to the right of the point (Figure 6.13). When there is conflict with other map features in the top right, try to move it above or clockwise to the right of the target. Only when there are no other nearby features that could be confused as the target, should you move the label to the least ideal position, such as to the left and underneath. Figure 6.14 is an excerpt of a map of upstate New York, designed to serve as a reference to primary school students. In the map, several lakes are labeled, but the lakes are small enough that they are essentially treated as points. As you examine the map, notice the preferential placement of labels of the small finger lakes, starting with the top and right side of the lakes where there was space, and moving around it as space allows.

In situations where there are multiple features and it is not immediately clear what the target feature is, **leadering** can be employed. Leadering refers to any line that is added to connect the label with its target feature. Leadering is used in Figure 6.14 to connect the labels for Lake Seneca and Lake Skaneateles to the corresponding lakes. Adding lines like this is sometimes necessary but should generally be avoided, particularly when there are lots of other line features on the map. Too much use of leadering can make the map complex and difficult to read.

FIGURE 6.13
Placement of labels for point symbols.

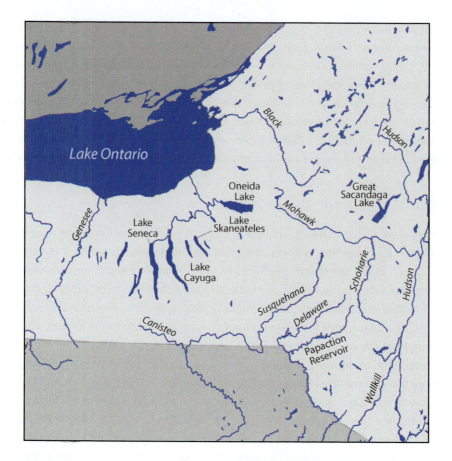

FIGURE 6.14

An excerpt of a basic map of lakes and rivers in upstate New York state, USA, to demonstrate label placement.

An additional solution for complicated maps is **key numbering**, the use of a number (or some other succinct symbol on the map, such as a letter) with a listed and ordered reference outside of the map. Figure A.7, a map of the buildings at Oxford University, employs this method. The building names (such as the "School of Geography and the Environment") are too long and cumbersome to effectively fit on the map. The readers can scan the list underneath the map to find the name of the building they are seeking and then locate the associated number on the map. The most ideal practice is to place the numbers in some ordered form on the map so that the reader can identify a pattern that guides him or her to the appropriate location (Freeman 2007)—the number "7" should be close to "8," for example. While the use of key numbering demands significant scanning and seeking work from the reader, maps of highly dense and complicated features may leave few other options.

Labels for line features should be placed centrally along the feature and should never be oriented to force the reader to read from right to left or bottom to top. *Adobe Illustrator*

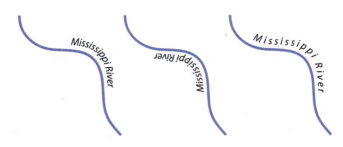

FIGURE 6.15
An example of splined text applied to mapping labels.

and mapping applications often enable you to place text along curves, useful for labeling sinuous features, such as rivers. The practice of placing text along a curved baseline is often referred to as **text splining**. "Splining" refers to a mathematical equation that defines curves between points. Due to the way that splining text affects the spacing between the letters, you should adjust the font and letter spacing appropriately so that the individual letters do not touch, leaving enough space around them so that they are easily legible. If the feature is so curvy that it starts to run the text in a right-left orientation or the spacing between the labels becomes very awkward, you should consider other ways of labeling the feature.

Figure 6.15 shows examples of splined text applied to a river symbol. The first example, on the left, shows the default spacing; when you add curvature to text, it often needs some additional tracking—space between the individual letters. The middle example shows an obvious blunder with direction; never make your readers read from right to left or from bottom to top (the example does both). The example on the right is the best application of splined text, oriented in the right direction with additional tracking space.

Area features can offer an ideal solution for label placement if the feature offers enough space; put the label right of the middle of the area! A common and useful technique in cartography is to add some tracking to help the label cover the extent of the feature being labeled. On the map of Europe in 1911 in the map gallery (Figure A.4), the label for the Atlantic Ocean uses significant tracking to convey the idea that the ocean extends throughout the entire body of water on the west side of the map. Similar techniques are employed throughout the map to convey the extents of countries and other features that would be otherwise difficult to show (notice the label for "Ottoman Empire," for instance). The technique is effective because the text remains easy to read and the letter spacing helps to convey important spatial information, such as the extent.

Summary

The goal of this chapter was to provide you with the concepts behind typography as well as some general guidelines for the effective use of text in maps. If you follow the key principles for the use of text, both as labels and other text elements in the map, you can leverage text to communicate valuable information succinctly and clearly to your audience. Text should be legible and should not interfere with other graphical elements in the layout. The size and position of the text should contribute to the intended visual hierarchy of the map, and labels should be clearly and unambiguously associated with the target features. Fonts, typefaces, and other features of the text should be leveraged to communicate the message and should not distract the reader. Many of these guidelines should appeal to your common sense, and at the end of the day, as always, you can learn a lot by thinking about what is effective for yourself and the readers of your map.

Discussion questions

1. Take some time to be mindful about the typography you observe in daily life. Where do you notice serif vs. non-serif text? Can you identify examples of decorative fonts? Can you find examples of typographic decisions that interfere with the legibility of text?
2. Find some examples of decorative fonts online or in some other medium. Write down a few words for the ideas that the font evokes a concept, idea, place, or period of history, and compare with the responses of your peers. Did the text evoke the same ideas? Does the history or use of those fonts prompt you to think of those ideas?
3. Examine the maps in the "Maps from the Wild" gallery or other maps for their use of typography. Which maps use text most effectively? Would including additional text or labels help any of the maps achieve its goals? Can you identify text you think the author should remove? Can you think of any typographic techniques the author could use to improve the map?
4. Think of an example of a map purpose and audience that you are interested in (or use an example provided by an instructor). What text should you include to help the map convey its message? What features would you label, and what features would you not label?

References

Adobe. 2018. "Glossary of Typographic Terms." Accessed February 15, 2020. https://www.adobe.com/products/type/adobe-type-references-tips/glossary.html.

Brewer, C.A. 2005. *Designing Better Maps: A Guide for GIS Users*. Redlands: ESRI Press.

Catich, E.M., and M.W. Gilroy. 1991. *The Origin of the Serif: Brush Writing & Roman Letters*. Catich Gallery, Davenport, IA: St. Ambrose University.

Freeman, H. 2005. "Automated Cartographic Text Placement." *Pattern Recognition Letters* 26 (3):287–297. doi: 10.1016/j.patrec.2004.10.023.

Freeman, Herbert. 2007. "On the Problem of Placing Names on a Map." *Journal of Visual Languages and Computing* 18 (5):458–469. doi: 10.1016/j.jvlc.2007.08.006.

Imhof, E. 1962. "Anordnung der Namen in der Karte." In *Annuaire International de Cartographie II*, 93–108. Zürich, Switzerland: Orell-Füssli Verlag.

Peterson, G.N. 2014. *GIS Cartography: A Guide to Effective Map Design, Second Edition*. Boca Raton, FL: CRC Press.

Sacks, D. 2003. *Language Visible: Unraveling the Mystery of the Alphabet from A to Z*. New York: Random House.

7 | Color in cartography

The decision about whether to produce a map in color or in grayscale was once often a critical one because color publications were much more expensive than black and white ones. Now that so much cartography is published in digital format, grayscale maps have become more of an exception than the standard. Maps that use color are generally capable of presenting a larger variety and richness of data than their black-and-white counterparts. If designed appropriately, color can also add aesthetic appeal and beauty to a map. Some caution is warranted, however. As with many of the tools in the cartographer's toolbox, color must be applied carefully, with prudent consideration of the goals and audience of the map.

This chapter begins by reviewing the fundamental concepts behind color and color theory, including the biological mechanisms behind the human color perception and the structure of different color models. It concludes with a discussion of the best practices for use of color in cartography, with guidelines for choosing colors that make maps accessible for color-deficient readers.

Color vision

Throughout our lives, we are immersed in electromagnetic radiation, the light-speed emission of energy as waves of subatomic particles. Radiation is typically categorized by its wavelength, which ranges from low-frequency and low-energy radio waves with large wavelengths that can be thousands of kilometers long, to high-energy gamma waves with frequencies less than ten picometers (a picometer is one-trillionth of a meter). Other types of radiation include x-rays, ultraviolet, and visible light. Human beings and most of the earth's animals have evolved to detect a convenient range of electromagnetic waves with frequencies between around 400 and 800 nanometers (nm), in the visible light spectrum.

Every object has physical properties that react to visible light in distinct ways. The **spectral signature** of a material is the combination of radiation waves that it reflects or

132

emits. Green grass, for example, absorbs much of radiation in the visible portion of the light spectrum but reflects radiation with frequencies between 520 and 560 nanometers, which we normally perceive as "green." The perception of "color" is our mind's interpretation of the combination of radio-magnetic frequencies reflected by an object and absorbed by special cells in our eyes.

Human eyes contain pigmented cells, called **cone cells**, that absorb visible light radiation and send signals to the brain in response. Working in tandem with cone cells are **rod cells**, which respond to the *intensity* of light signals. Rod cells are chiefly responsible for vision in low-light circumstances, and it is generally believed that they play little role in color vision perception.

There are two key theories about how this biological hardware functions. **Trichromatic theory** posits that there are three types of cone cells in the eyes (Young 1802). The theory suggests that cells detect different levels of radiation at different frequency ranges and the brain can process the combination of intensities of each signal, from which it yields the perception of color. **Opponent process theory** also works on the idea that the eye contains three different types of cells that receive and respond to specific ranges of visible light frequencies, but posits that the key information to the brain is the *difference* in intensity between the signals rather than the combined responses of individual cell types (Hering 1878). The concept of "opponent processes" comes from the idea that there are three opponent channels: red-green, blue-yellow, and black-white. The brain then processes the relative differences in intensity across the channels.

A portion of the population is **color vision deficient**, with an inability to distinguish between certain shades of color, more commonly referred to as "color blindness." A prominent theory of the mechanism behind color vision deficiency is that one or more of the classes of cones is deficient or lacks pigmentation. Color vision deficiency is usually inherited (Kalloniatis and Luu 2005). Most people who are color vision deficient normally *can* perceive color but have an impaired ability to distinguish between some colors and shades.

Because the genes that drive pigmentation in cone cells are on a sex-linked chromosome, men are much more likely (around 8% of the general population) than women (0.5%) to have some form of congenital color vision deficiency (Kalloniatis and Luu 2005). There are several defined forms of color vision deficiency, but by far the most common type is called **deuteranopia** ("red-green color blindness"), defined as an inability to distinguish between colors in the green, yellow, and red portions of the visible spectrum. Some research suggests that a minority of individuals (but perhaps as many as 12% of women) have a fourth class of cone pigmentation, which may give them "super color vision," or tetrachromacy (Jordan and Mollon 1993). The fourth cone cell type could enable these individuals to see millions of colors that most people are not able to perceive; however, the scientific evidence remains unclear on this topic (Jordan et al. 2010).

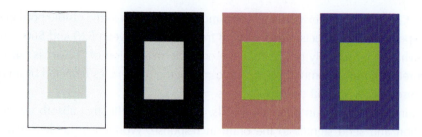

FIGURE 7.1
A demonstration of simultaneous contrast.

Critical to understanding the human perception of color is the idea that color is an entirely *subjective* experience. Individuals experience colors in different ways, resulting from their personal psychology and cultural experiences. Research shows, for example, that there is degradation in the blue-yellow channel with age, and that elderly people are less able to distinguish between colors than younger people (Schneck et al. 2014). Particularly when making design choices about color, it is important to remember that how we perceive colors is deeply influenced by context (arguably true for just about any human experience). **Simultaneous contrast** refers to the way that the perception of colors is affected by the presence of other colors around it, a concept introduced by a French chemist, Michel Chevreul, in the nineteenth century (Chevreul and Martel 1860). The color "blue" appears differently if it is surrounded by black, white, or a color hue, such as green or yellow. Figure 7.1 is an example of color combinations that often produce simultaneous contrast—the gray in the middle of the two boxes is precisely the same color, but they probably appear to you as different shades of gray due to the context given by the surrounding boxes.

Color terminology

Not only do color preferences differ across cultures (Taylor, Clifford, and Franklin 2013), but different languages have identified and named different colors altogether (Malt and Majid 2013). Consequently, there are a variety of ways to categorize, approach, and systematically describe the collection of electromagnetic frequencies that we call color. In color theory, this general term "color" includes its three primary facets: hue, saturation, and lightness.

Color hue is what most people think of when they hear the word "color" and includes all the colors of the visible spectrum (i.e., the "rainbow"). There is some disagreement around which specific colors are truly "unique," but there is universal consensus about the perception of colors green, red, yellow, and blue as **primary colors** (Miyahara 2003). Some color theories distinguish these as "psychologically primary" hues as the main hues that cone cells can distinguish and that most people recognize. Naturally, we tend to think of more than three or four colors (we perceive the world as being colored by a near infinite variety of hues, tones, and shades of color), but many named color hues, such as "purple" or "orange," are essentially perceptions our brain builds by interpreting the relative intensities of the primary or unique colors.

Another facet of color is its **saturation**, also called "intensity" or "purity." Our brains perceive white or gray as a combination of the main primary colors; an object that emits or reflects roughly equal amounts of radiation across the visible spectrum is perceived as white or gray. A "pure" form of red, on the other hand, is perceived from a *relatively* great amount of red frequency color radiation in combination with low amounts of the other colors. A useful way to think about saturation is the "vividness" of color; bright, vibrant colors are highly saturated. In contrast, a fully desaturated photograph or graphic appears as grayscale. Figure 7.2 displays columns of color patches showing different hues of color in descending order of saturation. Notice that they all converge to gray as saturation is decreased.

Brightness refers to the amount of radiation emitted from luminous objects, such as a computer screen, and **lightness** is the amount of reflected light from objects, such as the pages of a book. The brightness or lightness of color is commonly broadly referred to as **value**, particularly in computer applications. It is easy to conceptualize color value by imagining a color scheme in which both saturation and hue are held constant, such as a spectrum of grayscale that moves from white to gray to black. Figure 7.3 shows different color hues in descending order of lightness or value. Notice that the colors converge upon black as lightness decreases. Because our eyes have evolved to naturally adjust to different lighting situations, the value of colors is heavily influenced by the process of simultaneous contrast; colors appear as much darker when set against high value colors, but the same color appears lighter when set against low value colors.

FIGURE 7.2
Columns of color patches showing different hues in descending order of saturation.

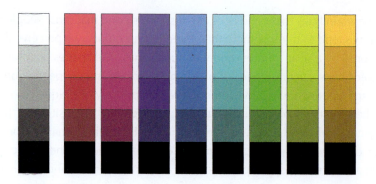

FIGURE 7.3
Different hues (in addition to a column of fully desaturated gray) in descending order of lightness or value.

Color models

Using these basic facets of color, we can create a systematic and objective way to define, describe, and communicate different colors for printing and digital displays. There are several color models designed for different coloration processes.

Subtractive color theory and the CMYK color model

The most critical distinction for understanding color models is whether color is produced by the *reflection* of light from the sun or another source, or by *luminance*, the emission of light from the color source itself. Throughout most of human history, colors were manufactured by combining different inks that absorb and reflect selected portions of the visible light spectrum. If you have ever tried mixing a variety of color paints, you will have noticed that it makes the paint darker rather than lighter (often culminating in a muddy brown color). Production of color by combining different colored inks is called **subtractive color theory**, built upon the process of selectively subtracting portions of the spectrum by adding inks. Subtractive color theory is based upon the **subtractive primary colors**, which are cyan, magenta, and yellow (Figure 7.4). Black is also included in print work, referred to as the **key color**.

Printers and printing presses produce the perception of color by combining different percentages of ink work with the **CMYK** (cyan, magenta, yellow, and key) color model. The model operates by specifying what percentage each of the components is used to produce any specific color. For example, the color white is produced by setting each of the components of a CMYK color model to zero. A dark purple color has a high percentage of cyan and magenta ink, along with some key ink. Figure 7.5 shows different combinations of cyan, magenta, and yellow ink. The percentage of each ink is decreased with distance from each ink's respective corner. Notice that the highlighted color patch

FIGURE 7.4
Subtractive colors.

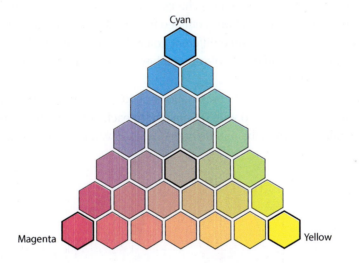

FIGURE 7.5
Combinations of cyan, magenta, and yellow ink.

in the center of the diagram, which contains 50% of each ink, appears as a mid-scale gray color. None of the color patches in the diagram include the key (black) ink.

Additive color theory and the RGB color model

The emergence of televisions and digital displays meant a new color model had to be devised because these devices *emit*, rather than reflect, visible light. Color mixing that is based on adding luminance of different colors is sensibly called **additive color theory**.

The main additive color model used in computer programming is called the **RGB color model** (RGB refers to "red," "green," and "blue"). The colors red, green, and blue are the building blocks of the **additive primary colors**. In contrast to subtractive color models, adding color makes the color brighter and more intense. Mixing colors with an additive model *adds* light, and so mixing colors produces lighter, brighter, whiter colors (Figure 7.6). All colors you see on a computer screen are produced by a combination of intensities from the red, green, and blue channels.

Each of the color channels is reported on a scale of intensity, ranging from 0 to 255, a total of 256 distinctions (counting zero along with the other 255 numbered values). This number undoubtedly emerged to accommodate eight-bit processing; eight bits yield 256 distinctions (consider the fact that $2^8 = 256$). An RGB value of 0,0,0 indicates that each of three color diodes in a pixel are set at zero intensity—essentially turned off—yielding the appearance of black. "Pure red" is made by setting the red diode to full intensity and turning off the blue and green channels (with RGB values of 255,0,0). Adding intensity from other channels changes the color and generally makes it appear brighter, since more light radiation is emitted from each pixel. Because the RGB model contains 256 distinctions of three different colors in combinations, it can yield a total of 167,772,216 (256^3) different colors, well beyond the number of colors any human being is able to distinguish.

Figure 7.7 shows different combinations of RGB intensities. The value for each channel intensity is decreased with distance from each color channel's respective corner. The highlighted cell in the middle has equal RGB values, set to 85 each. Figure 7.8

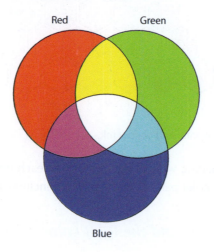

FIGURE 7.6
Additive colors.

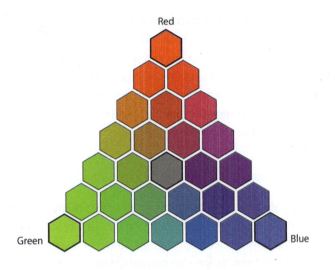

FIGURE 7.7
Combinations of RGB intensities.

	No Blue	Mid Blue	High Blue	No Blue	Mid Blue	High Blue	No Blue	Mid Blue	High Blue
High Green →	R = 0 / G = 255 / B = 0 / 00FF00	R = 0 / G = 255 / B = 119 / 00FF77	R = 0 / G = 255 / B = 255 / 00FFFF	R = 119 / G = 255 / B = 0 / 77FF00	R = 119 / G = 255 / B = 119 / 7FFF77	R = 119 / G = 255 / B = 255 / 77FFFF	R = 255 / G = 255 / B = 0 / FFFF00	R = 255 / G = 255 / B = 119 / FFFF77	R = 255 / G = 255 / B = 255 / FFFFFF
Mid Green →	R = 0 / G = 119 / B = 0 / 007700	R = 0 / G = 119 / B = 119 / 007777	R = 0 / G = 119 / B = 255 / 0077FF	R = 119 / G = 119 / B = 0 / 777700	R = 119 / G = 119 / B = 119 / 777777	R = 119 / G = 119 / B = 255 / 7777FF	R = 255 / G = 119 / B = 0 / FF7700	R = 255 / G = 119 / B = 119 / FF7777	R = 255 / G = 119 / B = 255 / FF77FF
No Green →	R = 0 / G = 0 / B = 0 / 000000	R = 0 / G = 0 / B = 119 / 000077	R = 0 / G = 0 / B = 255 / 0000FF	R = 119 / G = 0 / B = 0 / 770000	R = 119 / G = 0 / B = 119 / 770077	R = 119 / G = 0 / B = 255 / 7700FF	R = 255 / G = 0 / B = 0 / FF0000	R = 255 / G = 0 / B = 119 / FF0077	R = 255 / G = 0 / B = 255 / FF00FF
	No Red			**Mid Red**			**High Red**		

FIGURE 7.8
An alternative demonstration of RGB combinations that combine all the intensities of each of the channels.

is an alternative demonstration of RGB color combinations that combine all of the intensities of each color channel. Notice that the patch in the top right of the diagram, where all three channels are at full intensity, appears white. The patch on the lower left, where all channels are set to zero, appears black. The figure also displays the hexadecimal values for the colors.

While the ultimate appearance of color on a computer monitor, mobile phone, television, or other form of digital media varies by the hardware and by the specific device itself, a convenient application of the RGB color model is that it is easy for computer programmers and others to record, communicate, and reproduce specific colors. If you are exploring a map image and note a color or color scheme that works well, for example, you can record and use the RGB values for your own work. To make programming more convenient and better align the underlying architecture of computers, colors are often given in **hexadecimal**, a base 16 numerical system. This system is discussed in greater detail in Chapter 11.

Perceptual color theory and the HSV color models

One of the earliest systematic color systems, from which many of the subsequent color systems were derived, originated in 1907 from American scholar Albert Munsell (1907). The system was originally conceived through study of visual responses in humans (Cleland 2013). In the system, colors are comprised of three key values that match the way people think about color: hue, value (lightness), and chroma (saturation). Working as an educator, Munsell sought a system that would serve a role similar to music notation, enabling the specific colors to be systematically notated and communicated (Figure 7.9). The notation system consists of hue, given as five principle hues, namely, red, yellow, green, blue, and purple, as well as five intermediate hues that comprise a

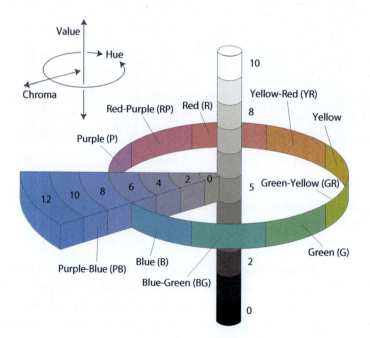

FIGURE 7.9
The Munsell Color System. Image adapted from Jacob Rus, 2007 (https://commons.wikimedia.org/wiki/File:Munsell-system.svg).

combination of the principles: YR (yellow-red), GY (green-yellow), BG (blue-green), PB (purple-blue), and RP (red-purple). Value is given on a scale of 0 (black) to 10 (white), and chroma is given a range of values from zero, representing "impure" colors with low saturation, to a range of higher values (the highest chroma value depends on the hue-chroma combination; some combinations have higher maximum chromas than others). Munsell's system enjoyed widespread adoption by organizations and regulatory agencies throughout the twentieth century.

The final color model was developed by computer programmers to provide a more intuitive color model that similarly mimics how we perceive and talk about color. The perceptual color schemes avoid seemingly counterintuitive color combinations of the additive and subtractive color schemes (consider that in the additive color model, for example, adding red to green produces yellow). The components of the **HSV color model** are the aforementioned three basic color characteristics: hue, saturation, and value. The model and its variants chiefly operate by converting the HSV values, designed for humans, to RGB values, designed for computers. Variants to these models, HSB (hue, saturation, brightness) and HSL (hue, saturation, and lightless), also appear in a variety of applications, but they function in fundamentally the same way.

Because there is no truly intuitive way to think about color hue in terms of intensity (unless you are well trained in how color hue functions vis-à-vis the wavelengths of electromagnetic radiation), color hue is represented as a 360-degree color wheel and is given as "degrees" on it. Primary red is at 0 degree, green is at 120 degree, and blue is at 240 degree (Figure 7.10).

The other components of the HSV color model, saturation and value, are given as percentage values between 0 and 100. As noted earlier, saturation refers to the intensity of the color. The closer the saturation value is to 100, the more intense the color, and the closer to 0, the closer to grayscale. Value refers to the lightness or brightness. A value of 100 is pure white, values in between are varying intensities of the color, and 0 turns the color to black (Figure 7.11).

While these various color models may initially seem complex, you can base a lot of your color selection in cartography from existing resources and models. A good way to gain familiarity with the color models and how they function is to experiment with different color palettes online or in a graphics program.

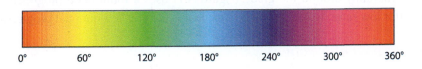

FIGURE 7.10
Color hues in the HSV spectrum.

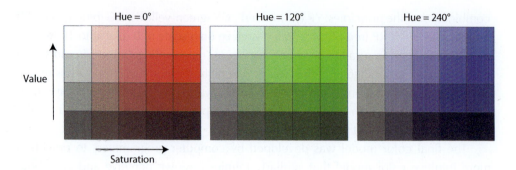

FIGURE 7.11
The HSV color grid.

Principles of color use in cartography

One of the most fundamental guidelines and conventions for using color to represent data in cartography comes from Bertin's ideas about visual variables (Bertin and Barbut 1967), discussed in Chapter 3. Colors that imply some order through saturation or lightness should be used for data that can be ordered, such as ordinal, interval, or ratio levels of measurement. Qualitative categories of data, such as "primary language spoken" or "type of forest," should use color in a way that does *not* imply any order, because the intent is to distinguish between categories, not to rank them.

Sequential color schemes

By the logic of Bertin's visual variables, color hue is a *qualitative* visual variable, while lightness/brightness/value (which I will refer to as "lightness" for brevity) and saturation are *quantitative* visual variables. The best solution for maps that show sequential data is usually to vary lightness and saturation so that there is a clear progression—from low intensity to high intensity or from dark to light (or both). If hue is varied only slightly or held constant, the ordering of the colors should be clear. People generally expect light, low-saturated values to represent lower data values than darker and more highly saturated colors, and so the best practice is to match your data symbolization to this expectation (Figure 7.12).

Small variations around the color wheel, using a progression of color hues that imply some order, can also effectively communicate ranked data (Figure 7.13). The map of child mortality around the globe (Figure A.1) uses a multi-hued progression from yellow (for low rates) to red (for high rates) that is clear as a sequential progression.

Qualitative color schemes

When you are communicating a qualitative distinction between map features, you should strive to keep saturation and lightness constant and vary the hue (Figure 7.14).

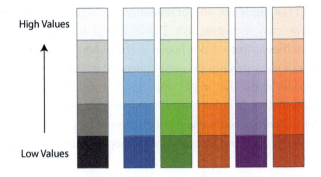

FIGURE 7.12
Color sequences designed for quantitative variables on maps. Image adapted from colorbrewer2.org.

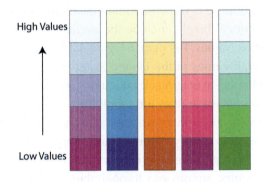

FIGURE 7.13
Examples of multi-hue color schemes. Image adapted from colorbrewer2.org.

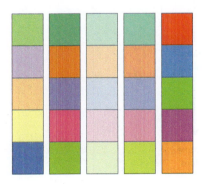

FIGURE 7.14
Color schemes suggested by colorbrewer2.org for qualitative maps.

The colors should not imply any sort of precedence or ranking, which can be misleading, and should be easy to distinguish. The historical map of Europe (Figure A.4) in the map gallery achieves this—color hue is used to represent the countries of Europe and none of the colors stands out from the others. Marc Monmonier (2018, 68) warns map readers to view maps that imply importance or preference of nominal categories through manipulation of color critically, as the cartographer may be attempting to guide the reader to some preconceived conclusion. If you do not wish to imply order, then you would expect no consistency in how people would order them based on the color.

Diverging color schemes

A **diverging color scheme** uses hue in combination with lightness and saturation to show sequential data (Figure 7.15). Contrasting hues indicate a **critical break** in the data, a point at which there is some important distinction between the data ranges. Good examples of critical breaks include the average value for the entire map, or negative versus positive values. The map in Figure A.11 uses a diverging color scheme to show cancer mortality rates in the United States. Health service areas with mortality rates close to the US average lack saturation, represented as a light gray color. Areas with higher-than-average rates are represented with a sequential brown color scheme, and those with lower-than-average rates are shown with shades of green. The effect is that the map clearly communicates important information about patterns: the areas with the highest relative cancer rates, clustered around the southeastern quarter of the country, are obvious through the contrasting color hue, while the

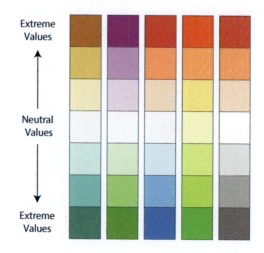

FIGURE 7.15
Diverging color schemes for quantitative data. Image adapted from colorbrewer2.org.

finer distinctions in the intensity of the data are communicated though lightness and saturation.

Communicating a lack of data with color

Another instance in which you can effectively use color hue is to represent areas without any data. Areas for which there are no data are indeed *qualitatively* distinguishable from areas that do. A generally good practice is to represent "no data" with a light gray color, as is done in Figure A.1 on the map showing child mortality rates by country. Gray-scale colors work well because the "no data" areas should not interfere with the interpretation of the patterns on the map. Gray is also generally a good color choice because you want the "no data" areas to fade into the back of the visual hierarchy—you do not want to draw the readers' attention to the parts of the map without data.

Respecting cartographic conventions and the culture of your audience

In some cases, cultural conventions guide readers to understand a palate of different hues as a clear sequential progression. A common and cartographically effective example of this can be observed in weather maps that appear on websites, newspapers, and newscasts. Since people normally associate the idea of "cold" with the color blue and the idea of "hot" with red, many weather maps are presented as a progression from blue to green, yellow, orange, and then red to represent increasingly warmer temperatures. Few readers have trouble making sense from these maps, and the color scheme ultimately serves its purpose well (see Figure A.9 for an example).

Most of the time when you select colors for your map, you can apply a healthy dose of common sense. In a map of a city, for example, what would you expect if you saw large patches of green in a city? Your first thought is probably that those patches are parks or open spaces. Even though the symbology may be clearly specified in the legend, a map that uses green to symbolize lakes or other bodies of water is likely to confuse or distract the map reader. Other common cultural conventions include symbolizing danger or negative attributes with red and, conversely, safe or positive features with green.

It is good practice to avoid using colors that are too intense or too highly saturated in any medium. You can reserve extremely intense colors for particularly important parts of the map, but too much color intensity overwhelms the map and quickly distracts readers from the message. Because color preference perception is deeply tied to cultural perceptions, it might be worth asking members of your map audience across different cultural groups to evaluate or interpret your color choices.

Making maps legible for color vision deficient readers

Making maps accessible to a broad audience can be challenging and often involves trade-offs. Producing tactile maps for blind map-readers, for instance, is seldom feasible for many mapping projects. There is a significant amount of research on making maps more accessible to readers, including for the visually impaired (e.g., Perkins 2002, Vrenko and Petrovic 2015). Designing a color scheme that is legible for color vision deficient readers is reasonably straightforward and can be achieved with little compromise or expense. As noted before, the most common form of color vision deficiency is *deuteranopia*, an inability to distinguish between shades of red and green. Figure 7.16 is a reworking of the RGB color grid presented in Figure 7.7, modified to show how the colors might appear to someone with deuteranopia. Color-deficient readers should have little trouble interpreting symbology on maps if it relies on the traditional quantitative visual variables of saturation and lightness. If you are including a map that uses color hue extensively, either to communicate some order in the data or to make important distinctions between qualitative features, you should avoid using red and green in combination.

If you want to see how a map might appear to someone with deuteranopia, you can test it on a website (such as hclwizard.org) that converts digital images so that you can see how they appear to someone who is color vision deficient. If you know someone who is color-blind (assuming that they do not mind, of course!), you can ask them a few guided questions about interpreting the data on the map to check whether they can read the key distinctions.

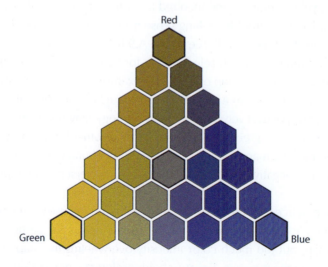

FIGURE 7.16
An RGB color grid after it has been transformed (via hclwizard.org) to mimic how it appears to someone with deuteranopia, the most prevalent form of color blindness.

Tools for color selection

There are several excellent tools online to help you explore colors and various color schemes for your maps. One of the best available resources frequently used by cartographers (which has certainly had an impact on the look of maps in the modern era) is an online tool called *ColorBrewer: Color Advice for Maps* (colorbrewer2.org). The concept of this tool was provided by Cynthia Brewer, a prominent cartographer and one of the world's leading experts on color use in cartography (e.g., Brewer 1996, Brewer et al. 1997, Olson and Brewer 1997, Brewer 2005). The tool enables users to select whether they wish to use a sequential, diverging, or qualitative color scheme; how many categories should be displayed; and for which medium (such as for printing, websites, etc.). Once the selections are made, the website displays a sample choropleth map using the colors as well as the color codes for the palette. While there are other online tools for building color palettes, *Color Brewer* is particularly useful because it is designed specifically to assist cartographers with their color choices.

Summary

Once you become aware of colors in mapping and graphical design and begin to note which color schemes work well together (and which do not), you can start to build your own set of color palettes for your cartographic work. When producing maps, you may discover that designing a color scheme with broad appeal is difficult. There are diverse and strong opinions about which color schemes are appealing or ugly; when you change the colors around to avoid what some readers perceive as unappealing, you may end up alienating other readers who would prefer the original option. Achieving this balance in cartographic color schemes requires appropriate application of visual variables to data, maintaining an awareness of how color combinations affect accessibility, and using colors in a way that builds upon the principles of cartographic design.

Discussion questions

1. Make a list of ten or so "main" colors, as you perceive them. Ask one of your peers to do the same, and then compare your responses. What differences do you notice and what do you think could explain them?
2. What colors do you or your peers like or dislike in maps?
3. What colors do you associate with specific ideas or emotions? Discuss the idea that some colors come "pre-loaded" with ideas or connotations. Can you think of differences in how different cultures perceive color?

4. Can you find examples of maps whose colors make the design unappealing or difficult to read? Try asking your peers if they agree and think about what you could do to improve the use of color in the maps.

References

Bertin, J., and M. Barbut. 1967. *Sémiologie graphique. Les Diagrammes, les Réseaux, les Cartes.* Paris; La Haye; Paris: Mouton; Gauthier-Villars.

Brewer, C.A. 1996. "Guidelines for Selecting Colors for Diverging Schemes on Maps." *Cartographic Journal* 33 (2):79–86. doi: 10.1179/caj.1996.33.2.79.

Brewer, C.A. 2005. *Designing Better Maps: A Guide for GIS Users.* Redlands, CA: ESRI Press.

Brewer, C.A., A.M. MacEachren, L.W. Pickle, and D. Herrmann. 1997. "Mapping Mortality: Evaluating Color Schemes for Choropleth Maps." *Annals of the Association of American Geographers* 87 (3):411–438. doi: 10.1111/1467–8306.00061.

Chevreul, M.E., and C. Martel. 1860. *The Principles of Harmony and Contrast of Colours, and Their Applications to the Arts: Including Painting, Interior Decoration, Tapestries, Carpets, Mosaics, Coloured Glazing, Paper-Staining, Calico-Printing, Letterpress Printing, Map-Colouring, Dress, Landscape and Flower Gardening, Etc.* London: Henry G. Bohn.

Cleland, T.M. 2013. *A Practical Description of the Munsell Color System and Suggestions for Its Use.* London: Literary Licensing, LLC.

Hering, E. 1878. *Zur Lehre vom Lichtsinne: Sechs Mittheilungen an die K. Akademie der Wissenschaften in Wien.* Wien: Druck Und Verlag Von Carl Gerold's Sohn.

Jordan, G., S.S. Deeb, J.M. Bosten, and J.D. Mollon. 2010. "The Dimensionality of Color Vision in Carriers of Anomalous Trichromacy." *Journal of Vision* 10 (8):19. doi: 10.1167/10.8.12.

Jordan, G., and J.D. Mollon. 1993. "A Study of Women Heterozygous for Colour Deficiencies." *Vision Research* 33 (11):1495–1508. doi: 10.1016/0042–6989(93)90143-k.

Kalloniatis, M., and C. Luu. 2005. "The Perception of Color." In *Webvision: The Organization of the Retina and Visual System*, edited by H. Kolb, E. Fernandez, and R. Nelson, 1–27. Salt Lake City: University of Utah Health Sciences Center.

Malt, B.C., and A. Majid. 2013. "How Thought Is Mapped into Words." *Wiley Interdisciplinary Reviews-Cognitive Science* 4 (6):583–597. doi: 10.1002/wcs.1251.

Miyahara, E. 2003. "Focal Colors and Unique Hues." *Perceptual and Motor Skills* 97 (3_suppl):1038–1042. doi: 10.2466/pms.2003.97.3f.1038.

Monmonier, M. 2018. *How to Lie with Maps, Third Edition.* Chicago, IL: University of Chicago Press.

Munsell, A.H. 1907. *A Color Notation.* New York: G. H. Ellis Company.

Olson, J.M., and C.A. Brewer. 1997. "An Evaluation of Color Selections to Accommodate Map Users with Color-Vision Impairments." *Annals of the Association of American Geographers* 87 (1):103–134.

Perkins, C. 2002. "Cartography: Progress in Tactile Mapping." *Progress in Human Geography* 26 (4):521–530. doi: 10.1191/0309132502ph383pr.

Schneck, M.E., G. Haegerstrom-Portnoy, L.A. Lott, and J.A. Brabyn. 2014. "Comparison of Panel D-15 Tests in a Large Older Population." *Optometry and Vision Science* 91 (3):284–290. doi: 10.1097/opx.0000000000000152.

Taylor, C., A. Clifford, and A. Franklin. 2013. "Color Preferences Are Not Universal." *Journal of Experimental Psychology-General* 142 (4):1015–1027. doi: 10.1037/a0030273.

Vrenko, D.Z., and D. Petrovic. 2015. "Effective Online Mapping and Map Viewer Design for the Senior Population." *Cartographic Journal* 52 (1):73–87. doi: 10.1179/1743277413y.0000000047.

Young, T. 1802. "II. The Bakerian Lecture. On the Theory of Light and Colours." *Philosophical Transactions of the Royal Society of London* 92:12–48. doi: 10.1098/rstl.1802.0004.

8 | 3D, animated, and web cartography

Students of cartography may appreciate the monumental and practically revolutionary pace with which the field has changed over the past several decades. Consider the changes in the work of the USGS, which provides comprehensive coverage of the United States with 1:24,000 scale topographic quadrangle maps. Each topographic map initially required significant manual labor involving several hundreds of hours of manual work. After surveyors performed the work on the ground, converting raw geographic data into maps involved scribing with manual etching on a light table, producing multiple exposures of the scribed sheets, and a painstaking editing and quality verification process that could involve additional field work in conjunction with aerial photography. In the early part of the twenty-first century, the USGS transformed their maps into digital formats, which can be much more rapidly edited, checked, and maintained on computers. Users can now download digital topographic maps from the USGS website immediately, rather than having to place a mail order for a poster-sized map sheet.

Since the 1980s, the video game industry has constantly pushed the standards for graphical displays, which, in turn, prompted the emergence of digital three-dimensional (3D) mapping, both for stereographic systems (a system that mimics real 3D viewing with two images from different perspectives) and for traditional displays that mimic 3D with dynamic displays. Two decades ago, producing 3D cartography was well beyond the resources of most cartographers. Anyone with a computer and the appropriate software and skills can now produce animated and 3D maps.

Digital technologies have also enabled the addition of the element of *time* to maps, giving birth to the widespread production of animated cartography. While the concept of animated cartography is not new—animated maps were initially produced in essentially the same way as an animated cartoon (see, for example, Thrower 1959)—the rise of computers and GIS cartography ushered an era in which they can be produced with ease. Following the development of the first digital map animations in the 1970s (Slocum et al. 2009), it is now possible to produce animated cartography with common geospatial applications, such as *ArcGIS*. The inclusion of these third and fourth

150

dimensions (time) to the mapping environment has meant that cartographers have needed to extend traditional theory to these frontiers of the field.

Few innovations have affected the study and practice of cartography as profoundly as the emergence of the Internet. The ability to host maps on the Internet significantly improved access and distribution of maps. The Internet has enabled spatial information to be managed, updated, and manipulated from almost anywhere. Web and mobile cartography encompass maps on computers, phones, and tablets. Maps can be connected to real-world phenomena to display spatial data dynamically, and some web services have enabled the visualization of near real-time geographic information. The emergence of web maps has greatly enhanced our ability to tailor geographic displays for specific purposes, creating *individualized* cartographic experiences. It has shifted the prevailing paradigm of cartography from a static to a much more flexible one.

While there are yet few universally accepted guidelines for these new forms of mapping, some work has explored best practices for mapping with these technologies; however, you can always bear fundamental questions for a mapping project in mind. The key principles of design in cartography apply to modern cartography as well as they did in the past; a basic feel for mapping and some common sense can get you a long way in working in new mapping media. This chapter provides an overview of the key terms, concepts, and some design principles for these modern forms of mapping.

Animated cartography

Animated cartography is the alteration of a map or map elements in real time. The purpose of animation is to draw attention to a component of the map or communicate thematic data. In the early 1990s, several cartographers identified important concepts in animated cartography (Dibiase et al. 1992, MacEachren 1995). The **duration** refers to the length of time a single frame is shown in an animation, similar to the concept of "frames per second" (FPS) in video animation. An animated visual variable can be the **rate of change**, or how much the characteristics of a feature—such as the location of an object or its size—changes within a given period. The duration and rates of change are related; an animation with a low duration and a high rate of change will produce smooth transitions, for instance. The **order** is the organizing framework in which an animated sequence of data is presented. The most natural idea that comes to mind when producing an animated map is a map **time series** that shows changes in data over time. However, animations may be presented in order of some other characteristic, such as one that shows the changing extents of forest fires across a progression of monthly precipitation rates.

In 1990, Phil Gersmehl wrote about a typology and an overview of the variety of tools and methods that can be employed in animated cartography through the use of metaphors, some of which I describe here. Readers are likely to be familiar with some examples of the types of animated cartography in the following discussion, which you can find by searching the Internet.

A **slideshow** is a simple progression of a sequence maps in a way that gives a user control of the map progression. A basic method for producing a slideshow is to build an ordered presentation of maps using software such as *PowerPoint*. The Environmental Systems Research Institute (ESRI) has incorporated a version of this type of animated cartography in their *Story Maps* software. Several examples are available for viewing at their website (https://storymaps.arcgis.com/en/app-list/map-series/).

Flipbooks are similar to slideshows insofar as they consist of a series of static maps, but they do not give the user control over the duration of the animation. Flipbooks function similarly to a cartoon or film, with a rapid succession of static maps to produce the illusion of movement and dynamic change. Many such maps are presented as animated gif images, which repeatedly cycle through a series of static images. The COVID-19 outbreak ushered in hundreds of examples of flipbook maps showing the progression of the virus through the population over time (try searching the Internet for "COVID-19 map animation" and you should easily find copious examples). Other common flipbook themes include maps of earthquakes, empires through history, and population change.

If the goal of the animation is to highlight a part of the map, rather than display an order or a progression of data, a **pointer** might be an appropriate tool. A pointer highlights some portion of the map by adding lines, illuminating a part of the layout, or adding notation to the map. One common use of pointers on maps is on newscasts in which the anchor marks a portion of the map during a presentation of a news event to highlight the location of a tornado touchdown or a hotspot in polling during a political election.

Metamorphosis occurs when the shape of a feature is changed, over time, in a map. Gersmehl (1990) provides the example of the shape of Greenland gradually changing shape and size to demonstrate the effect of different projections on angle and area. **Color cycling** is the rapid succession of colors along a path to simulate motion. This technique, common in computer graphics and on the Internet pages, can appear as waves of color or pattern across a surface, often constructed from a looped animation. Cartographers at Pennsylvania State University determined that color cycling serves as an effective tool for enhancing small features (Dibiase et al. 1992), drawing attention to a selection of features (Monmonier 1992), and conveying a sense of urgency to the reader (MacEachren 1995).

Also somewhat similar to the flipbook idea, the **stage and play** model can be used to animate or highlight thematic data. The concept behind the metaphor is that there is a static and unchanging "stage" to the map that forms the backdrop. The stage might include political boundaries, water bodies, or other contextual information. As in a theater, the "actors" are the components of the map that move around and interact with other map features. An excellent example of this model was produced by statistician and data visualization expert Nathan Yau (search the Internet for "Walmart animation by Nathan Yau"). The map shows a basic map of the United States, with the state boundaries and a few cities presented in dark tones in the background. The "actors" are points showing the appearance of new *Walmart* stores, a large department store chain in the USA, as well as appearance of its sister store, *Sam's Club*. As the years progress, over the span of seconds in the animated map, new stores are emphasized by a flash of a large colored circle that settles at a smaller size. As time progresses from the starting date of the map in 1965, the entire country is quickly inundated with over 4,000 points representing the stores. This example demonstrates the powerful story-telling potential behind animated cartography.

Another metaphor described in Germehl's paper is the use of a **model and camera** metaphor, which has become more technologically accessible since his article was originally published in 1990. The idea of this metaphor is that the user's perspective is placed in a virtual three-dimensional environment, which can be moved around the model. Thanks in part to the pressures on computer developers by the video gaming industry to develop realistic, first-person perspectives in dynamic virtual environments, this type of work is supported by a variety of software applications. ESRI, for example, supports map *scenes* that enables users to model 3D environments using map data. Modern graphics capabilities embellish surfaces with dynamic shadows and textures, effectively simulating movement through virtual 3D space. Users can build animations to give the perspective of "flying through" mapped data. You can find video examples of this work by searching for "ArcScene videos" or "flythrough maps" on the Internet. *Google Earth*, accessible from Google's Internet browser *Chrome*, is an excellent example of the application of this model.

Animations can improve cartographic comprehension by making data complexity more easily navigable or digestible for readers, and can include feedback mechanisms to help users locate information more quickly. The use of selective motion on map features has been demonstrated as an effective tool for drawing users' attention, though cartographers are cautioned to avoid overusing animation because it can become distracting and make the map more difficult to comprehend (Yilmaz et al. 2014). The transitions created through animation can ultimately reduce **cognitive load**, the amount of information users must hold in memory to make meaningful comparisons as well as reduce **cognitive lag**, the time it takes to mentally process a novel map interface. While research stemming from work in pedagogy has found that animations can improve comprehension of complex topics (Ng, Kalyuga, and Sweller 2013), it remains unclear how useful they are to improving mapping comprehension (Slocum et al. 2009). As a general principle, animations should be used strategically and are most effective when they are concise, unobtrusive, and aid in navigability.

Three-dimensional (3D) cartography

Prior to the development of advanced digital technologies, three-dimensional (3D) cartography was primarily under the domain of **tactile maps**, maps designed to be read through touch rather than vision. Tactile maps consisted of volumetric, 3D symbols, chiefly designed to make maps accessible to visually impaired people (Edman and Blind 1992). In a similar vein, **relief models** a physical 3D model of the environment with raised terrain, commonly found in museums (Figure 8.1).

Perceiving depth is one among the many incredible natural abilities our brains possess. Most of us can effortlessly gain a sense of how near or far the object is from our viewing perspective because our brains are well developed to compare the slightly different perspectives from each of our eyes to judge distance. For a flat image to be perceived as three-dimensional, a system must be in place that separates two distinct perspectives and prevents each eye from viewing the wrong image (i.e., the left eye should not view the right-eye image, and vice versa). Early stereophotographic viewers were developed in the middle of 1800s to produce the illusion of 3D photographs by separating two halves of a **stereopair** (a left- and right-eye image of a subject) with an optical lens.

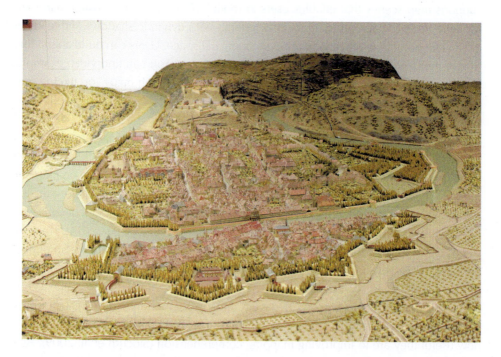

FIGURE 8.1
A relief map from the Musée du Temps de Besançon, Besançon, Franche-Comté, France.

Anaglyphs were developed around the 1950s to enable a single image to be viewed in three dimensions. One perspective is printed in shades of blue, with a second perspective superimposed over the first image in shades of red (Figure 8.2). Red-blue glasses serve to block the right-eye image from the left eye and vice versa, producing the illusion of a three-dimensional image.

The idea of imposing left- and right-eye perspectives onto a single plane in separable visual channels was later accomplished using polarized light. In the mid-1990s, several 3D "cave" projects began to develop three-dimensional maps and visualizations for geographic data. Over the course of the following years, affordable 3D visualization systems could be produced using a modern computer, two computer projectors, polarizing filters, and polarized glasses (Anthamatten and Ziegler 2006). Figure 8.3 shows an example of a 3D visualization system developed for use in university classrooms. The main components of the system include two projectors (one for each channel), polarizing filters, polarized glasses, and a computer capable of supporting multiple channels. Polarized light is filtered to only include light waves that move along a single plane. The right-eye and left-eye viewing channels can be separated with a polarized lens that only allows light on a particular plane to pass, and thus, polarized glasses can be used to filter out the wrong "channel." In the early 2000s, the film industry began producing 3D films with the technology.

ESRI released *ArcScene* in 2001, a plugin application designed specifically for showing spatial data in a 3D environment (the functionality of *ArcScene* has since been incorporated into the more recently released *ArcPro*). Using *ArcScene*, users can drape spatial data over a raster-based surface to indicate the base heights for maps and

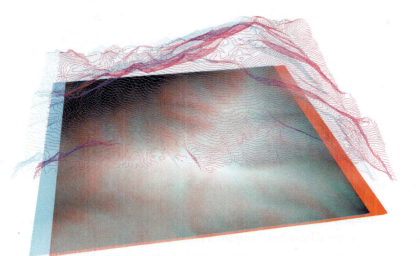

FIGURE 8.2
An example of a map anaglyph.

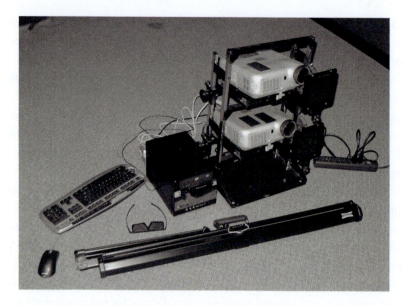

FIGURE 8.3

The hardware components of a 3D visualization system developed for university classrooms.

FIGURE 8.4

An example of an oblique, 3D perspective rendered by *ArcScene*.

mimic an oblique, 3D perspective. Figure 8.4 shows a 3D perspective map of contour lines that are assigned base heights from a raster data layer that contains elevation data. A georeferenced image of a topographic map is draped over the left half of the map. The software also enables users to manipulate the perspective, rotate the data, and build fly-through animations. Images that convey a 3D perspective on a flat screen are especially effective when combined with motion, a phenomenon called the **kinetic effect**. Traditional 2D imagery that conveys a 3D perspective without using any stereography is sometimes referred to as **2.5 D** or **pseudo-3D** maps.

While there is still promise behind the true 3D mapping representations, the inconveniences of setting up 3D mapping with multiple projectors and giving polarized glasses to viewers have hindered its widespread adoption. An additional drawback is that some viewers lack the ability to perceive depth, and many people experience visually induced motion sickness when they view stereographic images (Solimini 2013). True stereographic visualization is currently generated primarily in specialized visualization centers, universities, and museum exhibits. There is some growing interest in generating 3D **virtual reality** applications for cartography in fields such as education (Dos Santos 2018, Martinez-Grana et al. 2018) and planning (Valencia, Munoz-Nieto, and Rodriguez-Gonzalvez 2015, Martinez-Grana et al. 2018). The greatest impact of 3D visualization has been in computer-generated imagery that mimics 3D perspectives. Perhaps the use of true 3D cartography will expand as technology continues to develop cheaper and more convenient paths to stereographic presentation.

The introduction of computer-generated, pseudo-3D, and oblique perspectives has resulted in maps that are attractive and interesting to users that have found their way into a variety of digital and printed media (Häberling, Bär, and Hurni 2008). The ability to produce three-dimensional imagery provides a novel solution to an age-old problem in cartography—characterized and explored by twentieth-century cartographers such as Eduard Imhof (2007)—of how to approach the complex task of representing complex three-dimensional relief with two-dimensional symbology.

Several Swiss cartographers studied mapping with 3D perspectives (Häberling, Bär, and Hurni 2008). In their investigation, which consisted of interviews with expert map users who were given the chance to review several examples of cartography rendered in 3D, the authors proposed several principles of design for 3D maps. Their principles highlight some of the decisions that cartographers must grapple with as new technology is introduced. The authors found, for example, that a light application of haze effect or **atmospheric attenuation**, the simulation of atmospheric haze, as one might view over a large area of land area (such as from on top of a mountain), can serve to improve depth perception in a pseudo-3D image. They also recommend that users employ a **camera aspect** with a 45 degree angle, among other ideas about symbols, the sizes of map objects, and coloration.

A powerful and common technique for producing a 3D effect is to introduce subtle shadow effects or **hillshading** (Figure 8.5). Hillshading is usually performed with software to add a semi-transparent layer of shading to mimic shadows that would appear on undulations in the topography, usually by simulating a light source at the top left side of the map. This cartographic technique is extremely effective because it can efficiently communicate elevation without cluttering the map or interfering with other features.

157

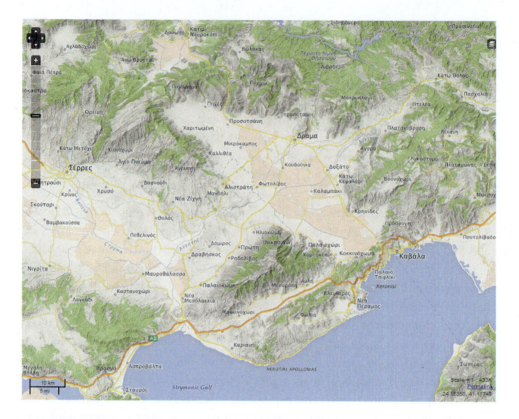

FIGURE 8.5
A map excerpt showing hillshading.
Data source: Openstreet Wiki 2020.

Web and mobile cartography

Maps first appeared on the web with the advent of graphics-supporting browsers in 1993 (Peterson 2010, Roth 2015). Starting in 1993, static maps could be scanned and uploaded on the Internet (see, for example, the Perry-Casteñeda Library Map Collection at https://legacy.lib.utexas.edu/maps/). However, the scans were often of poor quality, particularly after printing, and the files were large and cumbersome. The introduction of the *Adobe Acrobat* plugin enabled PDF files to be conveniently viewed and printed, which facilitated the storage of text and lines as vectors, enabling the display of high-resolution images.

Web-based and open-source cartography developed quickly alongside the development of the Internet. *MapServer*, a free, open-source web server for maps, was released in 1994. Two years later, *GeoSystems* (now known as *MapQuest*) began making interactive street maps with navigation services. In the following decade, the adoption of open-source standard programming languages such as Scalable Vector Graphics (svg) and eXtensible Markup Language (xml), provided cartographers with the ability to create and disseminate map and data repositories. *Google Maps*, a dynamic, open-source, browser-friendly map, made its debut in 2005, and ESRI introduced *ArcGIS Explorer*, a free GIS data viewer, the following year.

The first iPhone was released in 2007, hailing the onset of the "smart phone revolution." In 2009, Google released free, turn-by-turn navigation services and users began to interact with mobile cartography on a continuous basis. Since then, many innovations have contributed to web cartography. One important innovation was **cloud computing**, the use of a remote server rather than a local computer, to perform computing functions, such as the storage, management, processing, and analysis of data. **Rich Internet applications** (RIAs) are web-based applications that work like desktop software. APIs provided the building blocks for **open-source** software that can be examined and altered by anyone with access to the Internet and appropriate computing power. The precipitous change in the accessibility of technologies has also resulted in the rise of "amateur cartography," referring to the idea that anyone with computer and Internet access, and not necessarily with any formal training in cartography, can produce professional looking maps (Wilmott 2012, Yilmaz et al. 2014).

The rise of web and mobile cartography has given birth to a host of new terms in cartography. A **static map** has a defined extent that cannot be navigated or altered. In contrast, a **slippy map** can be navigated by panning and changing the scale or "**zoom level**." **Panning** enables the user to change the extent of the map, altering the geographic area it shows. Static maps may be preferable when changing scale does not contribute to the usability of a map and when the information is limited to a fixed extent. Slippy maps may be preferable when the data occupy a large extent, when understanding is improved through enabling zooming or panning, when overlay maps change dynamically or unpredictably, or when the map is intended for general exploration.

As with traditional cartography, the designer of any web map should bear in mind the purpose of the map and the intended audience. However, web map design has extended concepts such as user interface (UI) and user experience (UX) to cartographic design. **User interface design** is concerned with layout, graphics, and more nuanced cartography, sounds, and haptics (the use of touch) (Roth 2015), and focuses more on the ease and enjoyment of interacting with and manipulating the maps (Hassenzahl and Tractinsky 2006). User interface design becomes especially important as the diversity of the intended audience broadens. These additional considerations add complexity to the communication of geographic information.

Fluid maps do not use neat lines to separate the map from map elements or text and often occupy the entire web page. **Compartmentalized maps** graphically separate the map from other cartographic elements. The decision between the two should be based on complexity and functionality. Fluid maps give the map more space on the screen, and most map elements can be integrated without distracting from the map. Compartmentalized maps are preferred when there is competition between the visual hierarchy of the map and its elements. Separating the features may allow users to fully engage with one facet of the map at a time, maximizing the use of complex map elements and functionality (Muehlenhaus 2013). Interactive maps often now include **widgets**, components that enable users to interact with the application.

Interactivity is the key feature that distinguishes web and mobile cartography from traditional forms of mapping. Web maps can enable users to modify the map in countless ways. Users can adjust the scale and extent they view; change the maps' projections; add, remove, and filter the data on the map; modify symbology, search, and retrieve details; perform geoprocessing; toggle additional tools; and communicate location and orientation. It is ultimately the cartographer's job to judge how much interactivity is useful and what elements can be distracting or limiting. Interactivity can augment the complexity of the data presented while diminishing the amount of time it takes to get information, leading to a perceived increase in productivity and decrease in workload (Roth 2015), particularly when users are given control over animations (Ng, Kalyuga, and Sweller 2013). Additionally, researchers have found that increasing interactivity can reduce errors in the retrieval of spatial information (Yilmaz et al. 2014).

Web maps can dynamically shift between different interfaces by providing the ability to **toggle** map features or widgets. In *Google Maps* (https://www.google.com/maps), for instance, the map elements (search bar, basemap layer controls, zoom, and location controls) sit on top of the layout. However, when the user searches for a location, a compartmentalized window appears on the side of the screen, populated with the attributes associated with that data point. Displaying attribute data too complex for a **popup window** is a good use of a compartmentalized window. Additionally, compartmentalization can be useful when there are too many choices to modify the overlay maps or to filter data. Examples of these practices can be examined in the USGS Earth Explorer (https://earthexplorer.usgs.gov/) and ArcGIS online (https://www.arcgis.com/home/webmap/viewer.html).

An important challenge for web and mobile cartographers is accommodating the variety of ways and the media upon which their map will be displayed. Computers and mobile devices have a range of **aspect ratios** (the relation between the width and height of the display) and **screen resolutions** (usually given in pixels-per-inch [PPI], the number of pixels in one square inch of space on the screen). Table 8.1 and Figure 8.6 show a snapshot of the range of screen sizes on popular mobile devices in 2020. A basic understanding of the aspect ratio of the maps should inform the design of the map and layout. Ideally, the layout elements should be given the flexibility to dynamically adjust to the size and position based on the aspect ratio and resolution of the device

TABLE 8.1 Physical dimensions, pixel dimensions, resolutions, and aspect ratios of several currently popular mobile computing devices

Model	Physical Dimensions (cm)	Pixel Dimensions	Pixels per Inch	Aspect Ratio
Mobile Phones				
Apple IPhone 7+	6.7 × 12.2	1,080 × 1,920	401	9:16
Samsung Galaxy S8+	7.6 × 13.7	2,960 × 1,440	326	5:9
Google Pixel	6.2 × 11.1	1,920 × 1,080	441	16:9
BlackBerry Passport	7.8 × 7.8	1,440 × 1,440	453	1:1
Tablets				
Apple IPad Pro	19.7 × 26.2	2,732 × 2,048	264	3:4
Microsoft Surface Pro	19.5 × 14.6	2,736 × 1,824	263	2:3
Amazon Fire	9.4 × 15.1	600 × 1,024	171	5:8
Laptops				
Dell XPS 13	16.6 × 15.1	1,920 × 1,080	276	9:16
Apple MacBook 12-inch	16.2 × 25.9	2,304 × 1,440	226	5:8
Acer Chromebook	15.4 × 24.6	1,366 × 768	135	16:9

This table shows the diversity of screen sizes across mobile phones, tablets, and laptops using models sold at the time of this writing. Designing maps for display on mobile devices requires a great deal of careful consideration of the ways the maps may eventually be displayed. Data were derived from screensiz.es.

on which it is viewed. Map elements with fixed sizes and positions may interfere with viewing the map when there are changes to the aspect ratio. On the other hand, if elements are specified relative to the display, they may become too small to read on smaller screens.

Cartographic design in interactive cartography

Web maps can be made to look like traditional forms of cartography and generally include four basic elements: (1) the basemap, (2) overlay maps (thematic or reference data), (3) map elements, and (4) functionality. When you design a web map, you should think about the minimum number of components you can include to effectively reach your audience without losing any part of the message. Consider whether

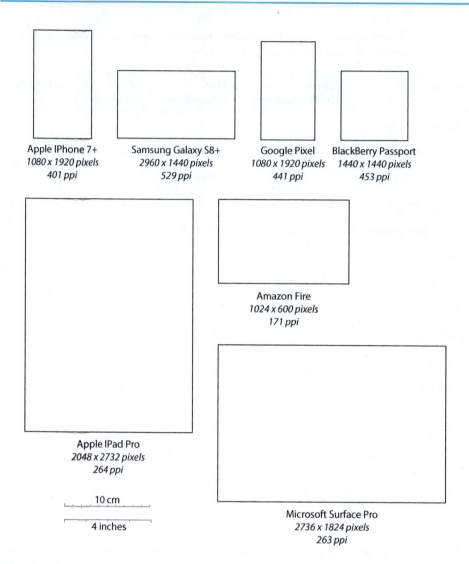

FIGURE 8.6

The physical sizes, pixel dimensions, and graphic resolutions (pixels per inch, PPI) of several mobile devices commonly in use at the time of this writing. Data derived from screensiz.es.

individual components improve readers' ability to comprehend the message; some components may not be necessary but can improve accessibility. Including redundant ways to achieve some task can improve the usability of the map.

Including conventional map elements can also enhance the readability of the map. Graphical zoom buttons, for example, have become conventional components of web maps, even though users tend to use a scroll wheel (on a mouse) or pinch (on a phone) to perform zoom functions. Finally, think about how components should be changed or modified based on the display devices you expect the audience to use.

Web cartography presents novel challenges for symbology, particularly in dynamic maps. Web maps can offer the ability to zoom, modify the transparency of symbols, or toggle multiple overlays. Zooming poses the question of which symbols should change based on the scale, how they should change, and whether they should be visible. It is important to avoid overlapping symbols or too much clutter. Symbols can be automatically removed at scales in which they contribute to cluttering and detract from legibility, then re-added as users zoom into the map.

Constructing a consistent visual hierarchy is often much more complex in web maps than in traditional maps. Dynamic elements that change size, shape, or color based on the data can affect the visual hierarchy. Some web maps enable the users to adjust the visual hierarchy to their preference by giving users the power to alter elements to emphasize specific map elements. The map can be designed so that the user can approach this manually, by dragging map elements to relocate them, for example, or changing the size of elements. This can also be performed automatically, through commands initiated with hovering, clicking, or tapping. Manual manipulations of visual hierarchy generally take more time to complete, which can make highly dynamic maps cumbersome to use. Ideally, however, the visual hierarchy should generally remain intact. Transitions should be designed to require minimal user input, be intuitive to the user, and not interfere with the user's comprehension of the map.

Changing symbology at different scales is a way to help users locate information without having to show it all at once. This is a form of **dynamic cartographic generalization**. A common example of this is clustering. At small scales, a polygon or single marker may represent multiple features. At large scales, these features can assume individual markers that would have otherwise cluttered the map at a small scale. This can also be achieved through the intensity and extent of symbology, as with a **heat map**. Dynamic heat maps show the density of information or attribute values through a gradient of different colors changing to fit the visible extent of the map.

Symbology in overlay maps can also be modified by allowing the user to change opacity, size, shape, or color of features. In maps where the information is complex, overlapping, or abundant, making the symbology interactive can allow users to create a custom cartographic experience tailored to their questions. This is often achieved through interactive legends with symbology options and filters.

Web and interactive maps also require special considerations for text. Reading font on a digital medium with backlighting, with a **variable resolution** that can be changed, affects some of the principles of typology. Changes in the image resolution can affect readability. For this reason, serif fonts are generally preferred over sans serif when you can anticipate resolution changes. The more simplistic san-serif fonts are more legible at low resolutions and less prone to distortion (Buckley 2012). The best web fonts have significant spacing between characters, and consistent line width (Muehlenhaus 2013), which aids in distinguishing between characters, mitigating variability in resolution, and making the text easier to read. Font size can change with resolution; at higher screen resolution, the text size can effectively become smaller. To work around this, some web cartographers design fonts for high resolutions. It is also possible to allow the user to control the font size. However, once again, providing undesirable levels of interactivity can jeopardize the visual hierarchy of labels.

Different devices and operating systems often have different fonts installed. Using an obscure font family may make the font unavailable to some of users. When a font is not available, a default is chosen by the machine, which can impact the legibility, placement, and size of the labels. It is a good idea to use common font families; examples include Verdana, Century Gothic, Arial, Helvetica, Trebuchet MS, Tahoma, Corbel, Myriad Pro, Myriad Web, Georgia, and Palatino (Buckley, 2012, Muehlenhaus, 2013).

Web maps can also have interactive text that serves as a button containing a popup, link, or animation. If you do not want the user to interact with the text, it is a good idea to make it un-selectable. In some situations, however, it can be advantageous to make text selectable so that the user can copy the text.

Color theory remains largely unchanged from traditional cartography. For the majority of web maps that are not designed to be printed at all, an additive color model, such as RBG (red, green, blue), is most applicable. Computer monitors function with an RGB model, which is the default for web images (Buckley 2012). Since additive and subtractive color spaces do not share all the same colors and monitors are configured for RGB, using an additive color model in web applications can create consistency across displays. Conventional norms for color symbology can also influence the map user experience; for instance, people are often accustomed to seeing major roads in color and minor roads in shades of gray or black. Following conventions can enable users to more quickly comprehend a map, focusing more on the message and less on deciphering symbology (Muehlenhaus 2013).

Summary

Producing successful cartography in these new media requires a broad set of technical skills. In addition to having some knowledge and experience of the topic that you are mapping, cartographers now make use of skills in programming, software development, web development, database management, video production, and 3D visualization, among others. Perhaps more than ever before, working in cartography requires learning and engagement with continually changing developments in the field. It is difficult to imagine where the field will head in the context of developments in artificial intelligence, 3D printing, and virtual reality, but new technologies certainly will take cartography to new frontiers that can be challenging to imagine.

While working in modern cartography adds a lot of additional decision-making to the process, the fundamental purposes of maps and the principles of design remain consistent. Fortunately, developments in cartography emerge upon a strong tradition of scholarly research. The next chapter provides an overview of the main themes in much of this ongoing work.

Discussion questions

1. What are some examples of mapping topics that are strengthened with animated or 3D cartography? How could these technologies bolster the scientific, political, or communication goals on these topics?
2. Examine two or three examples of map animations from the Internet. What is the key metaphor of the animations? Did the animation add to the potency or efficiency of the map's communication effort? If so, how or in what way?
3. In what ways do these technologies improve or harm the accessibility of maps? In what ways do they do this?
4. What are some of the drawbacks and challenges of the modern forms of cartography discussed in this chapter?

References

Anthamatten, P., and S.S. Ziegler. 2006. "Teaching Geography with 3-D Visualization Technology." *Journal of Geography* 105 (6):231–237. doi: 10.1080/00221340608978692.

Buckley, A. 2012. "Designing Great Web Maps." Accessed July 7, 2020. http://www.esri.com/news/arcuser/0612/designing-great-web-maps.html.

Dibiase, D., A.M. MacEachren, J.B. Krygier, and C. Reeves. 1992. "Animation and the Role of Map Design in Scientific Visualization." *Cartography and Geographic Information Systems* 19 (4):201–214. doi: 10.1559/152304092783721295.

Dos Santos, A.M.F. 2018. "(Web) Cartography and Augmented Reality: New Ways to Use Digital Technologies in Teaching Geography." *Geosaberes* 9 (17):14. doi: 10.26895/geosaberes.v9i17.647.

Edman, P., and American Foundation for the Blind. 1992. *Tactile Graphics*. New York: American Foundation for the Blind.

Gersmehl, P.J. 1990. "Choosing Tools: Nine Metaphors of Four-Dimensional Cartography." *Cartographic Perspectives* 5:3–15.

Häberling, C., H. Bär, and L. Hurni. 2008. "Proposed Dartographic Design Principles for 3D Maps: A Contribution to an Extended Cartographic Theory." *Cartographica: The International Journal for Geographic Information and Geovisualization* 43 (3):175–188. doi: 10.3138/carto.43.3.175.

Hassenzahl, M., and N. Tractinsky. 2006. "User Experience - A Research Agenda." *Behaviour & Information Technology* 25 (2):91–97. doi: 10.1080/01449290500330331.

Imhof, E. 2007. *Cartographic Relief Presentation*. Redlands, CA: ESRI Press.

MacEachren, A.M. 1995. *How Maps Work: Representation, Visualization, and Design*. New York: Guilford Press.

Martinez-Grana, A., J.A. Gonzalez-Delgado, C. Ramos, and J.C. Gonzalo. 2018. "Augmented Reality and Valorizing the Mesozoic Geological Heritage (Burgos, Spain)." *Sustainability* 10 (12):16. doi: 10.3390/su10124616.

Monmonier, M. 1992. "Authoring Graphic Scripts: Experiences and Principles." *Cartography and Geographic Information Systems* 19 (4):247–260. doi: 10.1559/152304092783721240.

Muehlenhaus, I. 2013. *Web Cartography: Map Design for Interactive and Mobile Devices*. Boca Raton, FL: CRC Press.

Ng, H.K., S. Kalyuga, and J. Sweller. 2013. "Reducing Transience during Animation: A Cognitive Load Perspective." *Educational Psychology* 33 (7):755–772. doi: 10.1080/01443410.2013.785050.

Openstreet Wiki. 2020. Accessed June 7, 2020. "File:OpenMapSurfer HillshadeLayer.png"

Peterson, M.P. 2010. *International Perspectives on Maps and the Internet*. Berlin, Heidelberg, Springer.

Roth, R.E. 2015. "Interactivity and Cartography: A Contemporary Perspective on User Interface and User Experience Design from Geospatial Professionals." *Cartographica: The International Journal for Geographic Information and Geovisualization* 50 (2):94–115. doi: 10.3138/cart.50.2.2427.

Slocum, T.A., R.B. McMaster, F.C. Kessler, and H.H. Howard. 2009. *Thematic Cartography and Geovisualization*. Upper Saddle River, NJ: Pearson Prentice Hall.

Solimini, A.G. 2013. "Are There Side Effects to Watching 3D Movies? A Prospective Crossover Observational Study on Visually Induced Motion Sickness." *PLOS One* 8 (2):8. doi: 10.1371/journal.pone.0056160.

Thrower, N. J.W. 1959. "Animated Cartography." *The Professional Geographer* 11 (6):9–12. doi: 10.1111/j.0033–0124.1959.116_9.x.

Valencia, J., A. Munoz-Nieto, and P. Rodriguez-Gonzalvez. 2015. "Virtual Modeling for Cities of the Future. State-of-the Art and Future Challenges." In *3d-Arch 2015-3d Virtual Reconstruction and Visualization of Complex Architectures*, edited by D. Gonzalez Aguilera, F. Remondino, J. Boehm, T. Kersten, and T. Fuse, 179–185. Gottingen: Copernicus GmbH.

Wilmott, C. 2012. "Living the Map: Cartographies of Mobile Media Environments." *Design Philosophy Papers* 10 (2):133–146. doi: 10.2752/089279312X13968781797878.

Yilmaz, A., S. Kemec, H. SebnemDuzgun, and M.P. Cakir. 2014. "Cognitive Aspects of Geovisualisation: A Case Study of Active and Passive Exploration in a Virtual Environment." In *Thematic Cartography for the Society*, edited by T. Bandrova, M. Konecny, and S. Zlatanova, 157–170. Cham: Springer International Publishing.

9 Scholarly research in cartography

In 1952, the well-known cartographer Arthur Robinson began his dissertation, *The Look of Maps*, with the words, "in the years to come it is likely that the twentieth century may be designated the golden era of cartography" (Robinson, Heap, and Roud 1952, 3). I believe he was correct, but this golden era has certainly extended through the first quarter of the twenty-first century. These are exciting times for scholarship in cartography. The ever-changing nature of computers and display technology, described in the previous chapter, has ensured that cartography scholars must constantly work to define new concepts and advance the theory and practice of the discipline.

The practice and study of cartography developed over millennia, when the idea of a map was constrained to a two-dimensional representation on a flat plane. The rise of global positioning systems, mobile technologies, and animated and three-dimensional cartography has required cartographers to adapt the principles of cartography for these new media and develop new ones. The open availability of data, and indeed the ability for users to contribute their own data for mapping, have fundamentally altered the purpose and social meaning of maps. Making, altering, or interacting with maps has become the domains of both professionals and amateurs, rather than of the few with substantial training in cartographic theory and practice.

While this book focuses primarily on the ideas behind good map design and the practical applications of mapmaking, no complete study of cartography should neglect the rich and diverse body of literature that explores maps from critical perspectives, so called because they challenge traditional power structures and paradigms of thinking. Maps are powerful tools and their creators have a responsibility to consider the implications of their work, including how they can be and often are used, intentionally or otherwise, to empower or disempower groups of people. Familiarity with this theoretical literature underscores the immense and sometimes subtle power of maps, which can engender an important perspective on ways maps affect our society, and which may prompt you to consider the meaning and context of your work. The first part of the chapter presents a review of literature exploring some of the modern trends in cartographic design theory and practice, and the second part introduces readers to concepts in critical cartography.

Current research on the practice of cartography

Scholarly research in cartography is part of a diverse and continually growing discipline that draws from a broad variety of fields. Important developments in technology, media, data availability, and society have introduced a broad range of problems for cartography, as well as opportunities for its advancement and growth.

It is beyond the scope of this book to provide a comprehensive review of the literature; the field is far too vast and dynamic for a single book chapter, and so I provide a summary overview of a few selected topics with examples of research from the last several decades to convey the nature and scope of the work. If you pursue a serious study of cartography, I recommend browsing through some of the key journals that publish work on research in cartography. These journals include most of the major geography journals (such as the *Annals of the American Association of Geographers*, *Professional Geographer*, and *Transactions of the Institute of British Geographers*), as well as several well-established journals dedicated to cartography and geographic information sciences. Some of these include the *Cartographic Journal*, *Cartography and Geographic Information Sciences*, and *Cartographica*, among many others. An assortment of other work can be found in various psychology, computer graphics, and social science journals.

Much scholarly work in the latter half of the twentieth century investigated cartographic design practices, such as which symbols or colors are the most effective at communicating complex data to novice users. Embedded in the modern manifestations of this first theme of research is the exploration of cartographic practices with special technological augmentation, such as the use of *sound* with mapping on mobile phones. A second theme discussed here is automated cartography, or "computational cartography," referring to the development of often complex computer models to assist with making cartographic decisions, such as how to best generalize features, place labels, and so forth. The next theme discusses the continual need to explore the cartographic applications of novel technologies and processes, such as open-source mapping. Finally, some of the literature discusses mapping applications for fields that are of special interest to professionals and cartographers working in specific areas, such as urban planning or natural resource management.

Studies of cartographic design

A body of research on *design principles* in cartography and the *usability* of maps continues to thrive in the twenty-first century. The specific purposes of this work are manifold and address a range of topics such as choosing evaluating visual variables, how readers interpret visual hierarchies, as well as design principles for modern forms of cartography, such as three-dimensional displays.

168

Much of this work is rooted in empirical research that draws from communications paradigms, a tradition ushered in by Arthur Robinson with his dissertation, published a few years after the Second World War (Robinson, Heap, and Roud 1952). Alan MacEachren summarizes this research tradition as based on the idea of "an information source tapped by a cartographer who determines what (and how) to depict, a map as the midpoint of the process, and a map user who 'reads' the map and develops some understanding of it by relating the map information to prior knowledge." (MacEachren 1995, 3). Figure 9.1 shows basic depiction of the communication paradigm of research in cartography. This body of research focuses on how to minimize information loss between the cartographer and the map reader by improving the map. While much of this body of work focuses on the ability of users to correctly extract knowledge from a map, much of it also examines user preferences in determining what practices are effective.

An excellent example of a map design study was led by Cynthia Brewer. Brewer reported on an experimental study to determine which color scheme was the most effective for enabling map readers to correctly answer questions about a map from the *Atlas of United States Mortality* (an example from this atlas is provided in Figure A.11) (Brewer et al. 1997). The authors asked study participants questions about different maps using a variety of color schemes and determined that maps that used diverging, color-bind safe color schemes yielded the most accurate results. Users expressed a preference for color schemes over grayscale, and they were able to read the maps most accurately in some multi-spectral schemes.

More recent work handles a broad range of cartographic design practices and often similarly includes some combination of quantitative (such as the formal evaluation of user knowledge following examination of a map) and qualitative research methods (such as the use of surveys to assess the preferences and attitudes of the map reader). Some examples of this work include a study to evaluate the effectiveness of cartograms and proposed best design practices (Nusrat and Kobourov 2016, Nusrat, Alam, and Kobourov 2018), the spacing, alignment, and design principles for legends (Li and Qin 2014, Golebiowska 2015, Qin and Li 2017), the use of point symbols to convey multiple types of information simultaneously (Klippel, Hardisty, and Weaver 2009), design principles for flow maps (Koylu and Guo 2017, Jenny et al. 2018), and design

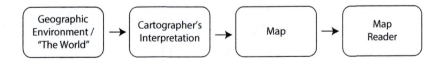

FIGURE 9.1
A basic depiction of the communication paradigm of research in cartography. This figure was adapted from MacEachren (1995, 4).

principles for animated (Cybulski 2016) or 3D maps (Lanca 1998, Hansen et al. 2004, Häberling, Bär, and Hurni 2008).

One sophisticated, empirical approach to research on map examines **foveation** or eye movement studies with trackers to collect data on where, in what order, and for how long users look at different parts of a map. In a study from Belgium, for example, a group of researchers compared the eye movements of novice and expert users on a series of topographic maps (Ooms, De Maeyer, and Fack 2014). Using data from eye trackers, they compared the viewing patterns of map readers and produced heat maps that depicted where participants looked at the map, showing the order in which they regarded the map elements and where they focused their visual attention. Figure 9.2 shows the paths of different users' scans of the maps. The authors concluded from this work that users generally spent most of their time fixing upon the reference frame, major roads, and rivers, perhaps underscoring the importance of those features in topographic maps. Other recent studies in cartography that use eye-movement tracking include evaluating the quality of urban plans (Burian, Popelka, and Beitlova 2018), examining the use of enhanced imagery base maps on the Internet (Dong et al. 2014), and exploring the use of crisis maps (Brychtova et al. 2013).

An important body of research investigates techniques and practices for making maps accessible to specific groups of individuals. The study of accessibility in cartography covers different population groups that include children (e.g., Hennerdal 2017)

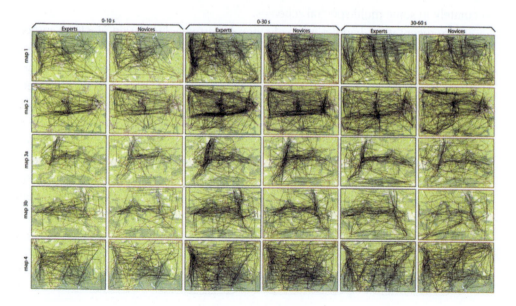

FIGURE 9.2

An example of a foveation analysis of topographic maps (Ooms, De Maeyer, and Fack 2014). Used with permission.

and elderly people (e.g., Vrenko and Petrovic 2015). Brewer has published extensively on good design practices for producing maps for color-blind audiences (Brewer 1996, Olson and Brewer 1997), and there are well-established guidelines for designing color-blind safe maps. More recent work has also explored mapping techniques for the visually impaired using **tactile mapping**—maps that are intended to be read and understood through touch (Perkins 2002, Rice et al. 2005, de Freitas 2012, Gual, Puyuelo, and Lloveras 2015)—as well as audio maps that use sound (Lin 2015).

Automated cartography

The introduction of digital forms of cartography in the second half of the twentieth century, along with the intense growth and improved availability of spatial data, has yielded the opportunity to support mapping with computer models. The availability of zettabytes of spatial data online has made the ability to rapidly produce map displays with computer technology critical. Computer software can assess objective information about a map, such as its scale or how close its features are to one another, to render on-the-fly decisions to display a readable map. **Automated cartography** has fueled the production of millions of maps in this way, to assist both with the *production* of carefully designed map products and the *display* of temporary or dynamic maps that appear on computer screens.

A fundamental problem with any initial attempt to work on traditionally human endeavors in a digital environment is the need to devise strict, operational, and objective rules. For much of its early development, cartographers were skeptical about the ability to truly produce automated cartography because computers are not (not *yet*, at least) capable of making subjective decisions (McMaster and Shea 1992). Eduard Imhof, a Swiss cartographer well known for his treatment of topography and study of the representation of elevation across complex landscapes, wrote that "the content and graphical structure of a complex, demanding map image can never be completely rendered in a completely automatic way" (Imhof 2007, 357). The last several decades have shown that, while it is arguably not yet possible to produce the highest quality maps through automation, it certainly has contributed in a substantial and meaningful way to modern map design and production.

One of the earliest initiatives in automated cartography was the development of line generalization algorithms. Building upon previous theoretical work in generalization (e.g., Tobler 1966), starting in the 1970s and 1980s, cartographers built objective rules and algorithms for generalizing point (McMaster and Shea 1992), line (Douglas and Peucker 1973, Jenks 1981), and area symbols (Monmonier 1983) in a way that could be implemented in digital cartography. Much of this early work is now commonly used in geospatial and cartography software. A key line simplification algorithm used in *ArcGIS*, for example, is based on work by Douglas and Peucker (1973). The algorithm accepts user inputs, such as how much displacement in the line is acceptable, and then removes points that make up the line if it does not cause the line to move further than a user-specified threshold.

Another common automated task in cartography is label placement. Particularly when users must deal with a complex series of labels, such as the names of streets in a road network or county labels for state maps, hundreds or thousands of labels need to be added to map in a way that is legible and clearly connected to the feature they are intended to represent. The labels

> should be such that no text obscures other text, there must be a clear association between the name and the feature to which it applies, there should be no ambiguity, and, for a map to be truly useful, the spatial relationships among the labeled features should be easy to comprehend.
>
> *(Freeman 2007, 458)*

Figure 9.3 shows some of Freeman's work on the correct placement of soil labels. Work to develop automated label placement algorithms has occurred for decades (e.g., Christensen, Marks, and Shieber 1995, Freeman 2005, Kakoulis and Tollis 2006, Wu et al. 2016, Marin and Pelegrin 2018). Recent work on this topic has explored more complex manifestations of modern graphical map displays, such as on interactive 3D maps (e.g., She et al. 2017).

Other key examples of automatic functions in cartography include selection (Buttenfield, Stanislawski, and Brewer 2011, Tinker et al. 2013), generalizing raster data

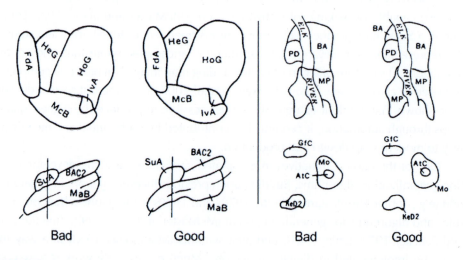

Bad Good Bad Good

FIGURE 9.3
Recommended placement for soil map labels using automated cartography (Freeman 2007). Used with permission.

(Schwarzbach et al. 2013), and automation of thematic mapping (Hey and Bill 2014, Arnold, Jenny, and White 2017). As computer technology becomes increasingly more sophisticated and the need for dynamic maps becomes commonplace, research on and development of automated cartography will continue to thrive.

Crowdsourced data and open-source cartography

The term **cybercartography**, or **web mapping**, first appeared during the rise of the Internet in 1997 at an International Cartographic Association conference (Taylor and Pyne 2010). The ideas presented described an emerging form of cartography that expanded mapping to cover a much wider range of topics than before, that would exist in a variety of multimedia formats, and that would be highly interactive, enabling a great deal of user-interaction with maps.

Since then, web mapping has flourished to encompass not only an extensive expansion of the kinds of maps being made and the people who make them, but also now includes the concept of **crowdsourced data** that are collected through the Internet by volunteers for public consumption. In a similar vein, **open-source mapping** refers to the idea that geospatial, mapping, and graphical software applications are built with computer code that is open to the public. Open-source software is not subjected to proprietary or commercial regulations, and so anyone can inspect, edit, or modify it. These developments have had an important impact on the way that cartographers approach maps altogether and have indelibly changed the role of maps in society in a broader sense—ultimately transforming the way governments function. In many ways, these developments have transformed cartography from a "top-down" field of work to a far more participatory one.

Crowdsourced mapping has received a great deal of attention in contemporary scholarly literature (Goodchild and Glennon 2010). Crowdsourced mapping is built upon **volunteered geographic information** (often referred to with its acronym, **VGI**), which is geographic or spatial data contributed by users to data collection platforms. Michael Goodchild (2007), who wrote about VGI during its inception, noted that some of the key benefits include its affordability, timeliness, and openness. One of the most successful VGI efforts has been *OpenStreetMap* (https://www.openstreetmap.org/) created in 2004 by a British entrepreneur. The site enables users from around the world to contribute spatial data from geographic positioning systems (GPS) units or personal observation, under an open-source license (Haklay and Weber 2008).

The emergence of VGI has led to a range of innovative and useful mapping applications. A prominent vein of research on crowdsourced data is its use in disaster relief efforts. Crowdsourced data can facilitate and maintain *ephemeral* data that change extremely rapidly, for which traditional cartographic data flows are far too slow. This advantage was demonstrated through an impressive effort in the aftermath of the

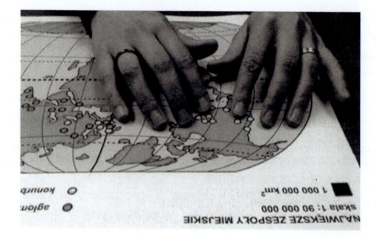

FIGURE 9.4
A user using interacting with tactile mapping in an open-source mapping project (Rice et al. 2013). Used with permission.

large-magnitude earthquake that struck Haiti in early 2010. The crisis in Haiti was compounded by the fact that there was little geospatial infrastructure to begin with; the lack of high-quality spatial information substantially hindered disaster relief efforts. Zook et al. (2010) describe the process and the role of crowdsourced mapping during the crisis, which was initiated by a flurry of social media activity. In a rather amazing, impromptu collaboration that was largely facilitated by *CrisisCommons*—an organization consisting of interested citizens, NGOs, government entities, and private companies for the purpose of coordinating volunteer technology support during disasters—VGI played an important and impactful role during the disaster. Users collected existing data from *OpenStreet Map* and other VGI sources to provide workers on the ground in Haiti with critical and reliable maps of the areas where they were working. Penn State has published a video series that describes some of the details of this effort (see geospatialrevolution.psu.edu).

Significant research has explored the cartographic applications of VGI. Rice et al. (2013) describe a crowdsourced mapping system to assist visually impaired people navigate the ever-changing obstacles in urban environments, such as construction sites (Figure 9.4). The combination of crowdsourced data and mobile phone technology enables maps that overcome some of the production and feasibility obstacles of traditional forms of mapping for the blind, such as tactile or haptic mapping.

Cartography for specialized topics

A final body of work explores mapping techniques and practices for specific disciplines, audiences, or topics, a body of literature that may have emerged from the

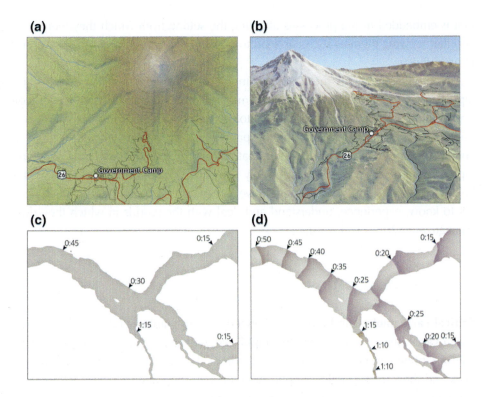

FIGURE 9.5
Maps examined in a study to evaluate the effectiveness of 3D perspective maps on user's ability to navigate with evacuation maps (Preppernau and Jenny 2015). Used with permission.

general expansion and democratization of cartography. Often this work starts by identifying a specific communication problem. Preppernau and Jenny (2015), for example, cite a general lack of understanding of geology and lack of experience with map use in the efforts to disseminate maps of volcanic hazards, essential for implementing successful mitigation and evacuation efforts (see Figure 9.5). They assembled a group of map readers and asked them to interpret terrain, estimate travel time, and select an evacuation route from available maps. They concluded from their study that users preferred 3D maps of volcanic hazards over 2D forms, and that the 3D maps were more effective at achieving the map goals.

There are copious examples of work over the last several decades on a broad range of topics. Examples include mapping climate and climate uncertainty (Kaye, Hartley, and Hemming 2012), seismic risk (Bostrom, Anselin, and Farris 2008), permafrost mapping (Heginbottom 2002), and indoor maps, such as maps of shopping malls (Nossum 2013). Scholarly disciplines outside of geography and cartography have also critically considered the role of maps and their potential to change the very approaches to the discipline. For example, Hacıgüzeller (2017) argues that maps can be studied as an act that is embedded in the processes affecting the setting from which they were borne, and that viewing them as such can lead to novel critical insight from the use of maps in archeology.

A critique emerged in the 1990s that much of the conventional literature in cartography is preoccupied with the form of a map (elements, symbolization, layout, color, design, etc.) and fails to consider the ontological work of maps in manifesting actual human landscapes. Works such as the paper by Hacıgüzeller tap into a large, nuanced, and complex body of literature that has emerged around mapping and cartography. Rather than serve as a purportedly "objective" model of reality, scholars have begun to critically consider "the generative role maps play in constructing how people get to know, experience, understand and deal with the worlds in which they dwell" (Hacıgüzeller 2017, 149–150).

Critical cartography

Critical cartography can be broadly defined as cartographic study which scrutinizes the linkages between mapping, geographic knowledge, and various forms of power. The discipline can be understood as a theoretical critique of existing mapping practices and their socio-political implications on the one hand, and as a technique for generating alternative, counter-hegemonic mapping practices on the other (Crampton and Krygier 2006). In critical cartography, maps are defined by the real-world implications of their use in social encounters. As such, critical cartography is not concerned with mapping practice or design per se, but rather with how maps reflect the social structures in which they are embedded, reinforce social or political power structures, or even serve to resist power. While study of critical cartography is not crucial for professional cartographers, it can prompt us to think about the meaning of our work: the context, impacts, and power of the maps we produce.

Many of the ideas in critical cartography draw from the seminal work of Michel Foucault, a French critical theorist who developed theories to explore the relation between knowledge and power. While most of Foucault's work focused on other social science disciplines, his work on social control was easily (perhaps even obviously) translated to cartography and geospatial technologies (Crampton 2009b). Almost all modern expressions of power involve mapping in some capacity because "spatiality" profoundly affects our social organization and interactions. Essential to the field is the recognition of the significance of cartography to a broad array of power structures that include political practice, imperialism, militarism, surveillance, news media, city planning, and natural resource management. Critical cartography further examines how maps mediate our individual interactions with our environment and how they may influence our perceptions of geographic space in problematic ways (Pickles 2004, Crampton 2009b).

The field of critical cartography is nuanced and complex. Bearing in mind that this summary ultimately simplifies much of this work, there are four related, but distinct, themes in the literature which define the field of critical cartography and connect the diverse perspectives therein. These perspectives view mapping as (1) an instrument of power, (2) the stage for the contestation of power, (3) an articulation of social meanings, and (4) a means of socio-material production and reproduction.

Mapping as an instrument of power

The fundamental insight of critical cartography is that all mapping practices are embedded in political and social contexts—cartographers are linked to the institutional settings in which their maps are commissioned. As such, every map is imbued with power. For example, governments and other powerful institutions require geographic knowledge (and representations of geographic space) to manage their territories and actualize their objectives. In this sense, maps are vital instruments of governance.

Much of the literature explores the historic uses of mapping technology to illustrate the efficacy of cartography for political practice, such as the role of surveying and mapping of land in the expression of colonial power (Wilford 2001). Invading governments leveraged maps to establish and defend territory, introduce land tenure and private property, design transportation and logistical networks, and extract natural resources (Akerman 2009, Edney 2009). Historically, maps were especially important for the construction of territorial space and national identities in Europe, eventually facilitating the emergence of the modern nation-state. A similar process unfolded in North America (Hannah 2000, Bryan and Wood 2015). Other work has explored the instrumental role of maps during the monarchies of medieval Europe (Buisseret 1992), the colonization of India (Edney 2009), the American bombing campaigns of the Second World War (Barnes 2008), and the creation of the state of Israel (Leuenberger and Schnell 2010) Figure 9.6 shows a map comparing Israel's size to the USA and the UK that appears in Leuenberger and Schnell's (2010) discussion of the role of maps in negotiating Israel's borders.

The emergence of digital technologies and satellite imagery has enhanced the ways in which mapping can be used as a tool of power (Monmonier 2004). Much of the literature discusses the use of geospatial technology for the surveillance of activists, migrants, and suspected criminals or terrorists (Amoore 2006). Pinder (2013) describes the mapping and policing of public space as a fundamental aspect of modern cities. The emergence of a surveillance state clearly illustrates how knowledge of space is necessary for the control of it, and that mapping can serve as an instrumental tool in the suppression, subjugation, and exploitation of others, which exemplifies the implementation of geographic knowledge as a mechanism of social control (Crampton and Krygier 2006, Crampton 2009a).

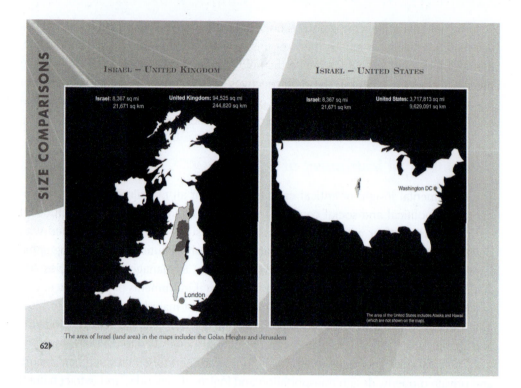

FIGURE 9.6
This is a map comparing Israel's size to the USA and the UK (Leuenberger and Schnell 2010). Used with permission.

Mapping as a stage for the contestation of power

A map can also serve to articulate and exercise power as well as to resist it (Bryan and Wood 2015). In this sense, cartographic representation embodies an active, multidimensional dialogue between conflicting visions of reality. This literature explores the ways in which normative maps are contested and how unconventional mapping practices can serve as a powerful form of resistance by producing alternative visions of past, present, and future geographies (Crampton 2009a). An example of a public effort to dispute official maps was a mapping initiative to display and publicize geographic patterns in airport noise and pollution (Cidell 2008, Meltzer 2017).

The key thread in this vein of critical cartography is **counter-mapping** and indigenous politics. Counter-mapping is a localized, participatory, and anti-technocratic form of cartographic practice which draws on existing experiential knowledges to generate alternative representations of geographic space (Bryan and Wood 2015). Just as maps have been instrumental in dispossessing indigenous peoples of their lands and livelihoods, mapping can also play a part in political and legal movements to reclaim spaces. Indigenous rights groups have used their own maps of historical and contemporary land use to successfully argue for legal revisions of land tenure which cede ownership back to the indigenous (Sparke 1995, Wainwright and Bryan 2009). Similarly, forest conservation zones in the postcolonial world have effectively excluded locals from forest areas and criminalized traditional forest uses. Some scholars argue that counter-mapping can serve as an effective strategy for regaining forest access and generating democratic dialogue regarding resource use (Harris and Hazen 2005).

The rise of highly interactive mapping and sources of volunteered geographic information has had an important impact on counter-mapping, resulting in a critical turn in political cartography. Crampton describes open source as the "undisciplining" or democratization of cartography, as non-experts now have access to creating maps and geospatial data (Crampton 2009a). Without the strictures of government, corporate, or academic cartography, mapmaking may be moving toward fundamentally different representational practices (Dodge and Kitchin 2013). Figure 9.7 shows an example of open-street mapping efforts of Bologna, Italy. The colors in the map represent different authors' contributions (Dodge and Kitchin 2013).

Maps as "text"; cartography as an articulation of social meaning

While cartography and map design have predominantly been discussed in this book as means of graphical communication of spatial information, they can also serve as a "text" that contains political, socio-cultural, or ethical meaning. This approach to critical cartography analyzes map design, layout, symbolization, and content for implied meaning about a culture's values and power structures. These can include subtle indications of social inequality such as race, gender, and, class, as well as nationalism or other ideologies. In many ways, this field is related to **discourse analysis** in the social sciences, which observes different forms of social discourse (such as from the popular media, speeches of politicians) to gain insight about the structures and underlying assumptions of society. The fundamental premise underlying this line of research in critical cartography is that the content and meaning of maps necessarily reflect the worldview of the individuals and institutions that produce them.

An important influence for this conceptualization of mapping is **semiotics**, or the study of social activities, practices, and processes that make use of signs and symbols in order to communicate and produce meaning (MacEachren 2004). Through this lens, maps are social significations which articulate propositions about the nature of places and spatial reality more generally. The notion that maps are a form of social discourse can be credited in large part to Denis Wood, whose popular book *The Power of Maps* was extremely influential in critical cartography (1992). For Wood, maps are never

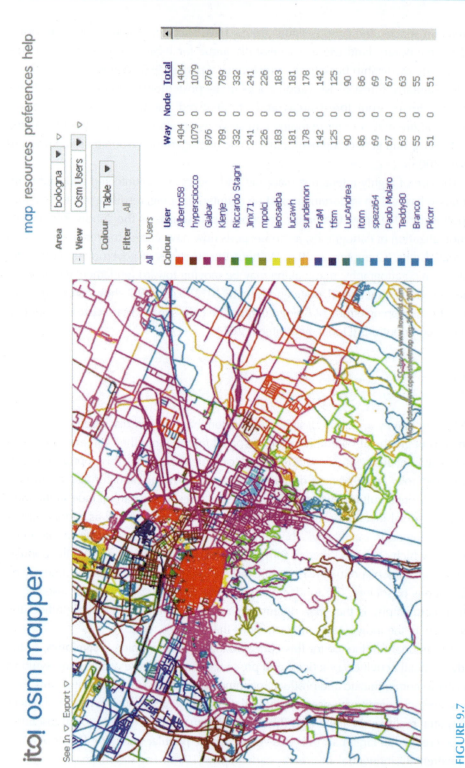

FIGURE 9.7
An example of open-street mapping efforts of Bologna, Italy (Dodge and Kitchin 2013). Used with permission.

simple representations of the world, but rather arguments about how the world is or should be. Their configuration is intrinsically *rhetorical* rather than scientific—they are instruments of communication and persuasion rather than depictions of an empirical reality (Wood and Fels 1992, Wood, Fels, and Krygier 2010). In line with the social theoretical traditions articulated by Foucault and others, Wood recognizes all modes of representation and abstraction as intrinsically political.

As I have emphasized throughout the first part of the book that the complexity of geographic space can never be represented in its entirety, cartographers must make highly selective decisions about what components of reality are worth representing. In this sense, how specific features are represented is often less telling than which features are omitted. This process of selection has powerful implications for how we ultimately perceive our world (Wood and Fels 1992).

A straightforward example of this perspective is how projections (such as the Mercator, a conformal projection that severely distorts area) can affect our sense of geopolitical space by enhancing some spaces, while diminishing others. Other works examine map layout elements, such as images and decoration. For example, Wood and Fels (1992) argue that the depictions of people on road trips on a road atlas of North Carolina reinforce white, heteronormative ideals of family and lifestyle.

Mapping and social production

Maps may not only influence political activity and how we perceive geographic space—they can directly shape our social reality by affecting how we interact with one another and with the environment. They can fundamentally affect political and economic decision-making as well as everyday social practices, both of which have *ontological* implications. The study of **ontology** concerns the nature of existence, but more specifically describes the process of how particular realities emerge: what interactions produce our world. Maps are critical instruments in the ontological process, and, as Pinder stresses, "maps are bound up with the production and reproduction of social life" (Pinder 1996, 405). In other words, cartography serves a role in creating as well as describing the world.

Many authors argue that maps are active; they are ultimately a material process and social practice rather than merely a model or representation of reality (Wood and Fels 1992, Crampton 2009a, Gustafson 2015). There is a distinction here between maps as objects and mapping as practice, which is akin to the philosophical distinction between *being* and *becoming*. "Being" or "map" defines a fixed or inert ontology, in which reality is composed of discrete, essential, and permanent elements. Conversely, "becoming" or "mapping" describes a lively, open ontology in which existence is constantly in motion, materializing and emerging each moment through dynamic processes (Crampton 2009b).

181

Some authors describe making and using maps as a form of "performativity" (Crampton 2009a, 2009b). Performativity describes speech acts or gestures which are "performed" in social space, thereby constructing and maintaining socio-material situations or a discursive practice that, through repetition and performance, can produce the phenomena it claims to only describe (Butler 2014, Glass and Rose-Redwood 2014). Mapmaking can be considered a speech act or discursive practice that "performs" a spatial reality, thus reifying that reality in social experiences.

Consequently, maps are ontogenetic, emerging spontaneously via social and discursive practice—they are active, mobile, and processual agents rather than static objects (Dodge and Kitchin 2013). As human renderings, maps are appendages of subjective intention and are always embedded in a living network of other social agents. In this sense, they participate in the social production of our reality, and cartography becomes more about the embodied activity of producing and using maps "in-the-world" than about maps themselves (Cosgrove 2005).

Summary

Scholarship in cartography spans a broad range of perspectives and disciplines, borrowing concepts from fields as diverse as communications, engineering, and social theory. Given the potential, power, and role of maps in our society, it is important that cartographers continue to be active and vigilant with respect to research in cartography. Future research will address novel challenges that encompass new technologies we may not be able to imagine now, as well as a society and culture that is changing at an accelerated pace.

Some attention to critical cartography is one among several important ways to keep abreast with developments in the field. While literature on critical cartography may not seem relevant to commercial and scientific cartographers who are often given specific, task-oriented jobs with clear objectives, familiarity with critical scholarship can remind us to think about the importance and meaning of our work.

Discussion questions

1. In what areas of work do you believe the field of cartography could benefit from more research on design practices? What kind of work on cartographic design are you interested in learning about as a means of improving your own cartographic endeavors?
2. Much of the discussion in the second half of the chapter was around the roles that maps play in society. What is your opinion on this? Do you believe that maps wield

power in some of the ways the authors under discussion describe? Can you identify specific examples of the power of maps?

3. What are some of the other ways that modern technologies and developments (such as the emergence of open-source mapping and sound-based mobile mapping) can be used to improve accessibility to maps? What are other ways maps can be used to address disability?

4. A critical development in cartography is the emergence of open-sourced and crowdsourced mapping. What are some of the benefits and drawbacks to this development? How could this other developments in the technology around cartography affect society in a broader sense?

References

Akerman, J.R. 2009. *The Imperial Map: Cartography and the Mastery of Empire*. Chicago, IL: University of Chicago Press.

Amoore, L. 2006. "Biometric Borders: Governing Mobilities in the War on Terror." *Political Geography* 25 (3):336–351. doi: 10.1016/j.polgeo.2006.02.001.

Arnold, N.D., B. Jenny, and D. White. 2017. "Automation and Evaluation of Graduated Dot Maps." *International Journal of Geographical Information Science* 31 (12):2524–2542. doi: 10.1080/13658816.2017.1359747.

Barnes, T.J. 2008. "Geography's Underworld: The Military-Industrial Complex, Mathematical Modelling and the Quantitative Revolution." *Geoforum* 39 (1):3–16. doi: 10.1016/j.geoforum.2007.09.006.

Bostrom, A., L. Anselin, and J. Farris. 2008. "Visualizing Seismic Risk and Uncertainty - A Review of Related Research." In *Strategies for Risk Communication: Evolution, Evidence, Experience*, edited by W.T. Tucker, S. Ferson, and A.M. Finkel, 29–40. Malden, MA: Wiley-Blackwell.

Brewer, C.A. 1996. "Guidelines for Selecting Colors for Diverging Schemes on Maps." *Cartographic Journal* 33 (2):79–86. doi: 10.1179/caj.1996.33.2.79.

Brewer, C.A., A.M. MacEachren, L.W. Pickle, and D. Herrmann. 1997. "Mapping Mortality: Evaluating Color Schemes for Choropleth Maps." *Annals of the Association of American Geographers* 87 (3):411–438. doi: 10.1111/1467-8306.00061.

Bryan, J., and D. Wood. 2015. *Weaponizing Maps: Indigenous Peoples and Counterinsurgency in the Americas*. New York: Guilford Press.

Brychtova, A., V. Paszto, L. Marek, J. Panek, and SGEM. 2013. "Web-Design Evaluation of the Crisis Map of the Czech Republic Using Eye-Tracking." In *Geoconference on Informatics, Geoinformatics and Remote Sensing – Conference Proceedings, Vol. I*, 1065–1072. Sofia: Stef92 Technology Ltd.

Buisseret, D., ed. 1992. *Monarchs, Ministers, and Maps: The Emergence of Cartography as a Tool of Government in Early Modern Europe*. Chicago, IL: University of Chicago Press.

Burian, J., S. Popelka, and M. Beitlova. 2018. "Evaluation of the Cartographical Quality of Urban Plans by Eye-Tracking." *ISPRS: International Journal of Geo-Information* 7 (5):25. doi: 10.3390/ijgi7050192.

Butler, J. 2014. *Bodies that Matter: On the Discursive Limits of "Sex"*. New York: Routledge.

Buttenfield, B.P., L.V. Stanislawski, and C.A. Brewer. 2011. "Adapting Generalization Tools to Physiographic Diversity for the United States National Hydrography Dataset." *Cartography and Geographic Information Science* 38 (3):289–301. doi: 10.1559/15230406382289.

Christensen, J., J. Marks, and S. Shieber. 1995. "An Empirical Study of Algorithms for Point-Feature Label Placement." *ACM Transactions on Graphics* 14 (3):203–232. doi: 10.1145/212332.212334.

Cidell, J. 2008. "Challenging the Contours: Critical Cartography, Local Knowledge, and the Public." *Environment and Planning A* 40 (5):1202–1218. doi: 10.1068/a38447.

Cosgrove, D. 2005. "Maps, Mapping, Modernity: Art and Cartography in the Twentieth Century." *Imago Mundi* 57 (1):35–54. doi: 10.2307/40233956.

Crampton, J.W. 2009a. "Cartography: Performative, Participatory, Political." *Progress in Human Geography* 33 (6):840–848. doi: 10.1177/0309132508105000.

Crampton, J.W. 2009b. *Mapping: A Critical Introduction to Cartography and GIS*. Hoboken, NJ: Wiley.

Crampton, J.W., and J. Krygier. 2006. "An Introduction to Critical Cartography." *ACME: An International Journal for Critical Geographies* 4 (1):1–33.

Cybulski, P. 2016. "Design Rules and Practices for Animated Maps Online." *Journal of Spatial Science* 61 (2):461–471. doi: 10.1080/14498596.2016.1147394.

de Freitas, M.I.C. 2012. "Tactile Cartography Experiences in Brazil: Methodologies and Didactical Material with Technological Resources for Inclusion of Blind and Low Vision Students." In *Inted2012: International Technology, Education and Development Conference*, edited by L.G. Chova, A.L. Martinez, and I.C. Torres, 1371–1379. Valenica: Iated-Int Assoc Technology Education and Development.

Dodge, M., and R. Kitchin. 2013. "Crowdsourced Cartography: Mapping Experience and Knowledge." *Environment and Planning A* 45 (1):19–36. doi: 10.1068/a44484.

Dong, W.H., H. Liao, R.E. Roth, and S.Y. Wang. 2014. "Eye Tracking to Explore the Potential of Enhanced Imagery Basemaps in Web Mapping." *Cartographic Journal* 51 (4):313–329. doi: 10.1179/1743277413y.0000000071.

Douglas, D., and T. Peucker. 1973. "Algorithms for the Reduction of the Number of Points Required to Represent a Digitized Line or Its Caricature." *Cartographica: The International Journal for Geographic Information and Geovisualization* 10 (2):112–122. doi: 10.3138/fm57-6770-u75u-7727.

Edney, M.H. 2009. *Mapping an Empire: The Geographical Construction of British India, 1765–1843*. Chicago, IL: University of Chicago Press.

Freeman, H. 2005. "Automated Cartographic Text Placement." *Pattern Recognition Letters* 26 (3):287–297. doi: 10.1016/j.patrec.2004.10.023.

Freeman, H. 2007. "On the Problem of Placing Names on a Map." *Journal of Visual Languages and Computing* 18 (5):458–469. doi: 10.1016/j.jvlc.2007.08.006.

Glass, M.R., and R. Rose-Redwood. 2014. *Performativity, Politics, and the Production of Social Space*. New York: Routledge.

Golebiowska, I. 2015. "Legend Layouts for Thematic Maps: A Case Study Integrating Usability Metrics with the Thinking Aloud Method." *Cartographic Journal* 52 (1):28–40. doi: 10.1179/1743277413y.0000000045.

Goodchild, M.F. 2007. "Citizens as Sensors: The World of Volunteered Geography." *GeoJournal* 69 (4):211–221. doi: 10.1007/s10708-007-9111-y.

Goodchild, M.F., and J.A. Glennon. 2010. "Crowdsourcing Geographic Information for Disaster Response: A Research Frontier." *International Journal of Digital Earth* 3 (3):231–241. doi: 10.1080/17538941003759255.

Gual, J., M. Puyuelo, and J. Lloveras. 2015. "The Effect of Volumetric (3D) Tactile Symbols within Inclusive Tactile Maps." *Applied Ergonomics* 48:1–10. doi: 10.1016/j.apergo.2014.10.018.

Gustafson, S. 2015. "Maps and Contradictions: Urban Political Ecology and Cartographic Expertise in Southern Appalachia." *Geoforum* 60:143–152. doi: 10.1016/j.geoforum.2015.01.017.

Häberling, C., H. Bär, and L. Hurni. 2008. "Proposed Dartographic Design Principles for 3D Maps: A Contribution to an Extended Cartographic Theory." *Cartographica: The International Journal for Geographic Information and Geovisualization* 43 (3):175–188. doi: 10.3138/carto.43.3.175.

Hacıgüzeller, P. 2017. "Archaeological (Digital) Maps as Performances: Towards Alternative Mappings." *Norwegian Archaeological Review* 50 (2):149–171. doi: 10.1080/00293652.2017.1393456.

Haklay, M., and P. Weber. 2008. "OpenStreetMap: User-Generated Street Maps." *Pervasive Computing* 7 (4):12–18. doi: 10.1109/mprv.2008.80.

Hannah, M.G. 2000. *Governmentality and the Mastery of Territory in Nineteenth-Century America*. Cambridge: Cambridge University Press.

Hansen, J.A., M. Barnett, J.G. MaKinster, and T. Keating. 2004. "The Impact of Three-Dimensional Computational Modeling on Student Understanding of Astronomy Concepts: A Aualitative Analysis." *International Journal of Science Education* 26 (13):1555–1575. doi: 10.1080/09500690420001673766.

Harris, H., and H. Hazen. 2005. "Power of Maps: (Counter) Mapping for Conservation." *ACME: An International Journal for Critical Geographies* 4 (1):99–130.

Heginbottom, J.A. 2002. "Permafrost Mapping: A Review." *Progress in Physical Geography* 26 (4):623–642. doi: 10.1191/0309133302pp355ra.

Hennerdal, P. 2017. "Continuity Markers as an Aid for Children in Finding the Peripheral Continuity of World Maps." *Cartography and Geographic Information Science* 44 (1):76–85. doi: 10.1080/15230406.2015.1109478.

Hey, A., and R. Bill. 2014. "Placing Dots in Dot Maps." *International Journal of Geographical Information Science* 28 (12):2417–2434. doi: 10.1080/13658816.2014.928822.

Imhof, E. 2007. *Cartographic Relief Presentation*. Redlands, CA: ESRI Press.

Jenks, G.F. 1981. "Lines, Computers, and Human Frailties." *Annals of the Association of American Geographers* 71 (1):1–10. doi: 10.1111/j.1467-8306.1981.tb01336.x.

Jenny, B., D.M. Stephen, I. Muehlenhaus, B.E. Marston, R. Sharma, E. Zhang, and H. Jenny. 2018. "Design Principles for Origin-Destination Flow Maps." *Cartography and Geographic Information Science* 45 (1):62–75. doi: 10.1080/15230406.2016.1262280.

Kakoulis, K.G., and I.G. Tollis. 2006. "Algorithms for the Multiple Label Placement Problem." *Computational Geometry-Theory and Applications* 35 (3):143–161. doi: 10.1016/j.comgeo.2006.03.005.

Kaye, N.R., A. Hartley, and D. Hemming. 2012. "Mapping the Climate: Guidance on Appropriate Techniques to Map Climate Variables and Their Uncertainty." *Geoscientific Model Development* 5 (1):245–256. doi: 10.5194/gmd-5-245-2012.

Klippel, A., F. Hardisty, and C. Weaver. 2009. "Star Plots: How Shape Characteristics Influence Classification Tasks." *Cartography and Geographic Information Science* 36 (2):149–163.

Koylu, C., and D.S. Guo. 2017. "Design and Evaluation of Line Symbolizations for Origin-Destination Flow Maps." *Information Visualization* 16 (4):309–331. doi: 10.1177/1473871616681375.

Lanca, M. 1998. "Three-Dimensional Representations of Contour Maps." *Contemporary Educational Psychology* 23 (1):22–41.

Leuenberger, C., and I. Schnell. 2010. "The Politics of Maps: Constructing National Territories in Israel." *Social Studies of Science* 40 (6):803–842. doi: 10.1177/0306312710370377.

Li, Z., and Z. Qin. 2014. "Spacing and Alignment Rules for Effective Legend Design." *Cartography and Geographic Information Science* 41 (4):348–362.

Lin, W. 2015. "The Hearing, the Mapping, and the Web: Investigating Emerging Online Sound Mapping Practices." *Landscape and Urban Planning* 142:187–197. doi: 10.1016/j.landurbplan.2015.08.007.

MacEachren, A.M. 1995. *How Maps work: Representation, Visualization, and Design*. New York: Guilford Press.

MacEachren, A.M. 2004. *How Maps Work: Representation, Visualization, and Design*. New York: Guilford Press.

Marin, A., and M. Pelegrin. 2018. "Towards Unambiguous Map Labeling - Integer Programming Approach and Heuristic Algorithm." *Expert Systems with Applications* 98:221–241. doi: 10.1016/j.eswa.2017.11.014.

McMaster, R.B., and K.S. Shea. 1992. *Generalization in Digital Cartography*. Washington, DC: Association of American Geographers.

Meltzer, E. 2017. "CDOT Did a New Air Auality Analysis for the I-70 Project, and Opponents Have Serious Questions about the Numbers." *Denverite*. https://denverite.com/2017/01/06/cdot-new-air-quality-analysis-70-project-project-opponents-think-numbers-look-fishy/.

Monmonier, M. 1983. "Raster-Mode Area Generalization for Land Use and Land Cover Maps." *Cartographica: The International Journal for Geographic Information and Geovisualization* 20 (4):65–91. doi: 10.3138/x572-0327-4670-1573.

Monmonier, M. 2004. *Spying with Maps: Surveillance Technologies and the Future of Privacy*. Chicago, IL: University of Chicago Press.

Nossum, A.S. 2013. "Developing a Framework for Describing and Comparing Indoor Maps." *Cartographic Journal* 50 (3):218–224. doi: 10.1179/1743277413y.0000000055.

Nusrat, S., M.J. Alam, and S. Kobourov. 2018. "Evaluating Cartogram Effectiveness." *IEEE Transactions on Visualization and Computer Graphics* 24 (2):1100–1113. doi: 10.1109/tvcg.2016.2642109.

Nusrat, S., and S. Kobourov. 2016. "The State of the Art in Cartograms." *Computer Graphics Forum* 35 (3):619–642. doi: 10.1111/cgf.12932.

Olson, J.M., and C.A. Brewer. 1997. "An Evaluation of Color Selections to Accommodate Map Users with Color-Vision Impairments." *Annals of the Association of American Geographers* 87 (1):103–134.

Ooms, K., P. de Maeyer, and V. Fack. 2014. "Study of the Attentive Behavior of Novice and Expert Map Users Using Eye Tracking." *Cartography and Geographic Information Science* 41 (1):37–54. doi: 10.1080/15230406.2013.860255.

Perkins, C. 2002. "Cartography: Progress in Tactile Mapping." *Progress in Human Geography* 26 (4):521–530. doi: 10.1191/0309132502ph383pr.

Pickles, J. 2004. *A History of Spaces: Cartographic Reason, Mapping, and the Geo-Coded World*. London: Routledge.

Pinder, D. 1996. "Subverting Cartography: The Situationists and Maps of the City." *Environment and Planning A* 28 (3):405–427. doi: 10.1068/a280405.

Pinder, D. 2013. *Visions of the City: Utopianism, Power and Politics in Twentieth Century Urbanism*. New York: Routledge.

Preppernau, C.A., and B. Jenny. 2015. "Three-Dimensional versus Conventional Volcanic Hazard Maps." *Natural Hazards* 78 (2):1329–1347. doi: 10.1007/s11069-015-1773-z.

Qin, Z., and Z.L. Li. 2017. "Grouping Rules for Effective Legend Design." *Cartographic Journal* 54 (1):36–47. doi: 10.1080/00087041.2016.1148105.

Rice, M.T., R.D. Jacobson, D.R. Caldwell, S.D. McDermott, F.I. Paez, A.O. Aburizaiza, K.M. Curtin, A. Stefanidis, and H. Qin. 2013. "Crowdsourcing Techniques for Augmenting Traditional Accessibility Maps with Transitory Obstacle Information." *Cartography and Geographic Information Science* 40 (3):210–219. doi: 10.1080/15230406.2013.799737.

Rice, M.T., R.D. Jacobson, R.G. Golledge, and D. Jones. 2005. "Design Considerations for Haptic and Auditory Map Interfaces." *Cartography and Geographic Information Science* 32 (4):381–391. doi: 10.1559/152304005775194656.

Robinson, A.H., H.R. Heap, and K. Roud. 1952. *The Look of Maps: An Examination of Cartographic Design*. Madison: University of Wisconsin Press.

Schwarzbach, F., J. Oksanen, L.T. Sarjakoski, and T. Sarjakoski. 2013. "From LiDAR Data to Forest Representation on Multi-Scale Maps." *Cartographic Journal* 50 (1):33–42. doi: 10.1179/1743277412y.0000000015.

She, J.F., J.L. Liu, C. Li, J.Q. Li, and Q.J. Wei. 2017. "A Line-Feature Label Placement Algorithm for Interactive 3D Map." *Computers & Graphics-Uk* 67:86–94. doi: 10.1016/j.cag.2017.06.002.

Sparke, M. 1995. "Between Demythologizing and Deconstructing the Map: Shawnadithit's New-Found-Land and the Alienation of Canada." *Cartographica: The International Journal for Geographic Information and Geovisualization* 32 (1):1–21. doi: 10.3138/ww47-6x0n-475q-7231.

Taylor, D.R.F., and S. Pyne. 2010. "The History and Development of the Theory and Practice of Cybercartography." *International Journal of Digital Earth* 3 (1):2–15. doi: 10.1080/17538940903155119.

Tinker, M., P. Anthamatten, J. Simley, and M.P. Finn. 2013. "A Method to Generalize Stream Flowlines in Small-Scale Maps by a Variable Flow-Based Pruning Threshold." *Cartography and Geographic Information Science* 40 (5):444–457. doi: 10.1080/15230406.2013.801701.

Tobler, W.R. 1966. *Numerical Map Generalization*. Department of Geography, University of Michigan. Ann Arbor: University of Michigan Press.

Vrenko, D.Z., and D. Petrovic. 2015. "Effective Online Mapping and Map Viewer Design for the Senior Population." *Cartographic Journal* 52 (1):73–87. doi: 10.1179/1743277413y.0000000047.

Wainwright, J., and J. Bryan. 2009. "Cartography, Territory, Property: Postcolonial Reflections on Indigenous Counter-Mapping in Nicaragua and Belize." *Cultural Geographies* 16 (2):153–178. doi: 10.1177/1474474008101515.

Wilford, J.N. 2001. *The Mapmakers*. New York: Vintage Books.

Wood, D., and J. Fels. 1992. *The Power of Maps*. New York: Guilford Press.

Wood, D., J. Fels, and J. Krygier. 2010. *Rethinking the Power of Maps*. New York: Guilford Press.

Wu, C.B., Y. Ding, X.X. Zhou, and G.N. Lu. 2016. "A Grid Algorithm Suitable for Line and Area Feature Label Placement." *Environmental Earth Sciences* 75 (20):11. doi: 10.1007/s12665-016-6190-4.

Zook, M., M. Graham, T. Shelton, and S. Gorman. 2010. "Volunteered Geographic Information and Crowdsourcing Disaster Relief: A Case Study of the Haitian Earthquake." *World Medical & Health Policy* 2 (2):7–33. doi: 10.2202/1948-4682.1069.

10 | Data in mapping

A groundbreaking development in the evolution of cartography is the increased availability of data. Obtaining even secondary data was once an arduous and time-consuming process. I can recall reading through tables of data in the 1990s, with a bag of dimes in hand for making Xerox copies at a library so that I later could spend hours manually typing them into a database. Now there is far more data immediately available on the Internet than we can effectively use. The emergence of volunteered geographic information and open sources of data has meant enormous growth of data, much of which are available to everyone with Internet access.

Data are at the very core of the power of cartography. Spatial data can be used to tell powerful stories, with the cartographer playing the role of a storyteller (Wood, Kaiser, and Abramms 2006, Koch 2016). People often view highly polished maps without considering the notion that the quality of map rests largely upon the quality of the data upon which it is based. A seemingly straightforward and accurate presentation of data can hide or distort data complexity. One of the chief responsibilities of a cartographer is to understand their data and present their story with honesty and clarity.

A key skill for many twenty-first-century professionals is the ability to find, critically evaluate, and collect data, and then put data to use to leverage its power in as honest way as possible. This chapter presents an overview of some of the concepts behind spatial data as well as practical ideas about collecting data and incorporating it into your maps. It concludes with a discussion of data certainty and some ideas about communicating data certainty in maps.

Data concepts and terminology

The most basic distinction in data used in mapping is between spatial and aspatial data. Spatial data, also commonly referred to as **georeferenced data**, are simply data with references to location, such as coordinate pairs. Most spatial data have associated

attribute data, or information about the feature. Aspatial data include any data that do not have explicit locations attached to them.

At the beginning of just about any cartography project, you need to find data for the **base map**, the features of the map that serve as the background and provide spatial context. Few maps make a lot of sense if they do not include some combination of political boundaries, rivers, bodies of water, cities, or road networks. GIS software vendors, government agencies (such as the USGS) and open-sourced GIS data repositories provide ample sources of freely available and high-quality digital base map data.

Spatial data can assume a variety of forms, but two key file systems continue to dominate the scene. **Shape files (*.shp)** were introduced in the early 1990s by ESRI, the company that produces *ArcGIS*. The file consists of a list of coordinates that are used to build the GIS "primitives"—points, lines, and areas. For the GIS to read a shape file, it requires multiple, additional files to function properly, which should be saved in the same directory as the SHP file. The shape index format file (.shx) contains the positional index of the SHP file, essentially serving to facilitate its processing and display. The database file (.dbf) contains the attribute data of the geometry. These three file types are required for any display of the geometries of the shape file, but other files are also often included. A projection file (.prj) contains the coordinate system information. The second major type of file is a **geodatabase (.gdb)** file, now the primary type of spatial data storage for *ArcGIS*. Geodatabase files can contain geographic feature classes, datasets, rasters, and tables. While GDB files are technically a collection of different types of files, they are often packaged together and can be downloaded as a single file.

While a huge amount of spatial data from traditional paper maps has already been converted into a form that can be used and manipulated by a GIS, you may find good information on a non-georeferenced digital image or on a paper map. In these cases, you may **digitize** the map from its paper form into your GIS or graphics program. Digitization is a form of data capture that involves finding or creating an image of map and then manually entering the data into a GIS by entering the locations of the points or tracing lines. A few decades ago, the ability to digitize spatial data was an important skill and a frequent task for cartographers, when the work required rather expensive and cumbersome **digitizing tables**, specially designed tables for digitizing work.

Digitizing is now performed on computer screens with digital images. If you locate a good digital image of a map online, you can load the image in your GIS and start the work. If you get your hands on a paper map, you can scan it to a digital raster image. With the digital image file on hand, then the next step is to perform **georeferencing** to link the image to real-world locations. After the image is georeferenced, the final step is to trace the features into the GIS. While the process of digitization can be time-consuming, the ability to incorporate data into your map from non-digital sources greatly expands the repertoire of data available for mapping with a GIS. Figure 10.1 shows some digitizing work of a schoolyard in Denver, based off of aerial imagery.

FIGURE 10.1
An example of some digitizing work in a GIS.

A good example of useful data in the form of static, ungeoreferenced maps is the Sanborn insurance maps (Figure 10.2). During the nineteenth and twentieth centuries, the Sanborn Maps Company conducted extensive mapping work of cities and towns around the United States to aid fire insurance companies with liability assessment. The maps are well-known for their reliability and accuracy, useful for researchers and cartographers interested in mapping the history and development of cities.

Ministries or departments of governments often publish a wealth of data that empower users to perform mapping work. A great deal of useful data is available in the form of aspatial, tabular data that cannot be immediately mapped in a GIS. Much of this data *are* referenced to location through the inclusion of the names of political or administrative units, such as cities, towns, provinces, or countries. Some of the data include latitude-longitude coordinates that can be directly transformed into spatial data in a GIS. Common file formats for tabular data include **CSV** (comma separated values), **DBF** (database files), or **XLS** (Microsoft Excel). CSV files, which are little more than text files with commas to separate rows and columns, are the simplest and most broadly operable type of aspatial data file.

Cartographers can link tabular data to locations by performing a **database join** on the data. A join is an automated process for matching a **key field** from separate tables to

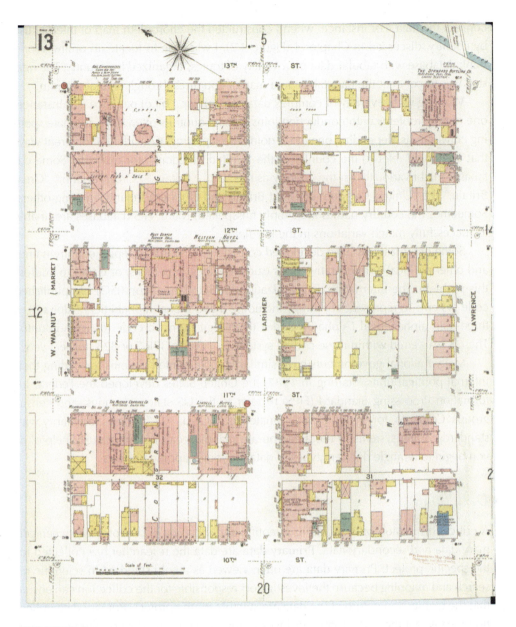

FIGURE 10.2
An example of a Sanborn insurance maps, a series originally produced to manage liability and risk for fire insurance companies (Sanborn Insurance Company 1904).

connect the data. For instance, when I constructed the choropleth map of Internet use by country, discussed in Chapter 4 (Figure 4.2), I was able to locate an online World Bank database with tabular data on a variety of topics, organized by country. The next step was to connect the data tables to a map of countries through a database join.

Working with a variety of complex data can call for a bit of patience. For instance, one of the problems with conducting joins is that countries often have alternate spelling conventions, depending on who is doing the reporting. Consider the different ways that one could refer to the country of the "United Kingdom," for example. Some databases might variously list the country as the "UK," "Britain," "Great_Britain," "Great Britain," the "United Kingdom of Great Britain and Northern Ireland," among others. Sometimes the country names are simply misspelled! Many GIS databases are unable to successfully match variations in the key fields.

Some organizations have developed consistent standards to report country names and other political-administrative units (such as states, provinces, or cities) to address this problem. The International Organization for Standardization (ISO) has published code systems to represent countries, including a two-digital alphabetical (Alpha-2), a three-digit alphabetical (Alpha-3), and a three-digit numeric code for each country. The UK is referred within these coding systems as "GB," "GBR," or 826, respectively.

Many national governments have similarly implemented systems for coding their internal political units. Starting in 1987, the United States developed the Federal Information Processing Standard (FIPS) to refer to the US states, counties, and census groups. Colorado, for example, is designated as "CO" or "08." A basic familiarity with standard coding systems of the area you are mapping can be enormously helpful as you begin to wrangle different types of data for your mapping projects.

Data quality

A critical distinction often made in scientific and mapping endeavors is between primary data and secondary data. **Primary data** are data the researcher has collected for the specific project. Primary data are often viewed as a gold standard in scientific research and mapping because the investigators responsible for the collection can ensure that the quality and nature of the data are suitable for the project. Primary data collection means that the cartographer or investigator can tailor the data on the question or topic being explored, and, as such, can enable the scientific question or the purpose of the project to drive focused data collection. Research questions that include primary data collection designed to address the specific question or project at hand can be termed **science-driven**, organized around the scientific question.

One of the major drawbacks of primary data is that data collection can be extremely expensive, time-consuming, and, in some cases, entirely infeasible. Nearly every modern cartographer must rely on **secondary data**, data collected by sources other than the author or investigator. Secondary data include government census reports, tabular data on political or administrative units, geographic boundary files, or any data collected by organizations or researchers for other purposes. Modern spatial data on coastlines, country borders, and other features are often built upon centuries of work that began with the initiation of careful surveying and geodetic measurements in the seventeenth century. Ultimately, most modern cartography stands on the shoulders of people throughout history who have brought us to the quality and quantity of cartographic data we have today.

One of the drawbacks of using secondary data is that it encourages **data-driven** work that is designed to accommodate available data. The analysis and the use of secondary data are valuable, given the gargantuan amount of data available and the human effort that has gone into its collection, but additional care must be taken to ensure that the methods are sound. The decisions about how secondary data were collected were made by people not involved with the work at hand, and so control of key decisions is beyond the domain of the cartographer or researcher. The *Atlas of Mortality*, for example (see Figure A.11), used death certificates to calculate age-adjusted mortality rates for various causes of death in the USA. Death certificates may seem like a highly reliable source of data to casual observers; however, a study of death certificates in 2002 revealed that over half of them did not contain properly completed cause-of-death statements and that nearly half of them disagreed with reports from full autopsies (Sehdev and Hutchins 2001). This part of the story can be lost in a polished map of mortality.

A problem facing modern geospatial data is **data error propagation**. When you download a GIS file of country borders to use in your map, think about how those data got from the real world to your GIS and, ultimately, to your final map. Once upon a time, some government authority negotiated, surveyed, and carefully mapped the borders. Others used the surveying work to construct authoritative paper maps to govern regions or drive policies. Someone then may have digitized the paper maps into a digital vector file that a GIS can read and use. The digital map probably underwent transformations due to changes in the coordinate and projection systems and may have undergone a process of generalization. Each time data are transformed or manipulated, there are multiple opportunities for the introduction of error into the data, by basic human error or some other means. The term "data error propagation" refers to the idea that the effects of error are intensified as a set of data passes through multiple users. It is a good idea to read about the **data lineage**—the history of the data from its original collection and its transformation along the way to its current form.

Spatial data can drive policies that have important, real-world implications, and measures of data quality can help to guide policy decisions. While few would suggest that cartographers should not use secondary data, it is the responsibility of the cartographer to appropriately communicate the limitations of secondary data in their mapping work. The quality of geospatial data can be communicated through **metadata**, or data about the data that should be included with the data files. In order to work with the problem of communicating data quality to its users, various organizations have set up data quality standards that should be reported in spatial metadata.

There are several components to spatial data quality that pertain to location, time, and theme of the data (Veregin 2005). Accuracy is the degree to which the data align with the reality. **Spatial accuracy** is the estimated agreement between the location of features on the map and their actual locations in the real world. Forms of spatial inaccuracy can include both vertical and horizontal displacement. **Temporal accuracy** refers to the "currentness" of the data, or how up-to-date the data are. In ideal situations, the time of the data is reported so that users can determine how current the data are (or how well the data work for a historical mapping project). **Thematic accuracy** refers to the accuracy of the attribute data, that is, the quality of information about the spatial features.

Precision, synonymous with the term "resolution," is the amount of detail that is reported in the data. As with the other concepts of spatial data quality, precision can be applied to location, time, or the theme of the data. In raster data, the spatial resolution refers to the size of the individual pixels, the smallest unit of discernable space in a raster data set. In vector data, the resolution is the smallest unit of distance reported. Temporal resolution similarly refers to the unit of time reported, and thematic resolution refers to the smallest discernable unit measured. As noted elsewhere, the two concepts are independent: precise data are not necessarily accurate.

Data consistency is the absence of conflicts in a database, referring to the logical integrity of the database architecture. While there are ways to apply the concept of data consistency to the temporal and thematic realms of data, it is most commonly applied to spatial components (Veregin 2005). An important component of data consistency is **topology**, a branch of mathematics dating back to the work of Euler. An example of data inconsistency is when borders between adjacent counties do not match up perfectly; a border between adjacent counties should, by definition, be defined by a single line. Modern GIS software includes mechanisms to identify and correct errors in topology.

Completeness is how well the objects in the database represent the universe it purports to represent. Data completeness can refer to errors of commission, the inclusion of non-existing features, as well as omission (Veregin 2005). If the data claim to contain all the pine trees in a golf course, for example, existing trees that are not included in the data can be considered an example of poor data completeness. An important part of the idea is that the features are correctly specified. If a database claims to contain information on federally funded hospitals for veterans, it is complete if it contains all the hospitals fitting that description, even if it excludes other types of hospitals.

Data certainty in mapping

The perception of maps as authoritative sources of data means that even basic and seemingly trivial decisions about symbolization can end up having meaningful real-world implications. An illustrative story about the power of maps comes from geographer Phil Gersmehl (1985), who describes his experience with designing maps of soils in the late 1970s. Gersmehl produced a series of point distribution maps of the major soil orders in the United States in 1977, for which he had to make some difficult decisions about symbology in his map. The Western US included a class of soils called "Histosols" that were scattered across the region and appear inconsistently, but were abundant when aggregated over a large area. He used point symbols to represent the distribution of soils in the United States on his map and so decided to use a single point to symbolize these small patches of histosols in the Western US. The footnote for the map included an explanation for the symbol: that a single point was used to represent soils that were scattered across a large region (Figure 10.3).

Some years later, Germehl encountered a map titled "Peatlands in the United States" (US Department of Energy 1979) as various organizations had become interested in harnessing peat as a form of renewable energy at the time. He noticed patches of peatlands in Utah and Colorado, which did not make sense, given his knowledge of soils in the USA. Curious about where the authors of the map came up with the idea

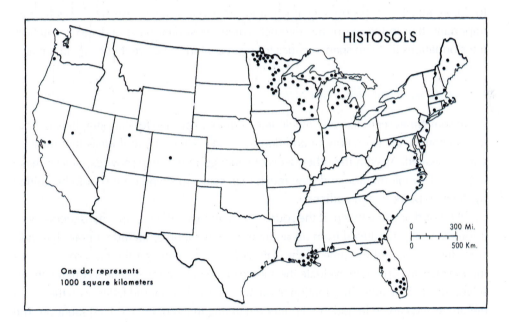

FIGURE 10.3
Phil Gersmehl's (1977) map of Histosols in the United States. Used with permission.

of showing peat resources in these western states, he investigated the lineage of the data and discovered that it could be traced back through several maps back to his own published work on soils in 1977!

Gersmehl suggests that a multimillion-dollar project to harness renewable energy may have been directed to Colorado, over a competing site in Massachusetts, because of the way that others had incorrectly interpreted his map. He concluded that he should have been clearer and more explicit about how he symbolized isolated patches of histosols across a large area, that he could have better explained what histosols are to an audience he suspected had poor background knowledge on soils science (particularly since people often confuse "Histosols" with "peat soils"), and that he could have better articulated how reliable the map actually was. Gersmehl (1985, 334) concludes from this experience that:

> a person who puts information on a map has a duty to be fair to the data, to be clear to the reader, and to try to anticipate the ways in which a third person may be affected by a foreseeable misinterpretation of the map. At the very least that third duty should include a resolute refusal to display or even imply any more accuracy and precision than we can justify.

Gersmehl's tale demonstrates the care with which representing data in cartography should be handled. The importance of thinking carefully about data quality and how to communicate the history, accuracy, and precision of data have never been more important than in the era of the Internet, where most data have undergone multiple transformations that can span decades (or even centuries) before it ends up in a map.

Communicating data certainty in maps

It can sometimes seem like an unwelcome additional burden for cartographers to have to also convey data certainty in addition to all the other data and ideas that must be communicated. As with the other decisions that go into a map, how explicit to make data certainty information and how much map space to spend on it should be guided by the map purpose and audience.

It is good practice to report the source of the mapped data on the maps themselves. If the data were produced from an original, primary source, include a note that indicates the owner and the time of the data collection. Including the data source gives map readers the option to evaluate the source of the data or investigate its quality themselves. Most of the time, information about the data source should fall low on the visual hierarchy and can be included without demanding much of the reader's attention. The source of the data can be listed in author-date form or simply as a URL that contains the original data files. If you include a URL, it is often a good idea to indicate the date along with the URL, since it may disappear in future. Make sure to avoid printing a URL that is password-protected!

When it is appropriate for the map, you can report additional information, such as how the data were collected (e.g., through surveys, direct observation, through a random or a stratified spatial sample). If you believe this information lacks relevance for the audience, you can omit details like this from the map itself. If, however, you anticipate that the mapped data will be used for scientific research or to guide policy decisions, it might be useful to include a short paragraph of text on the map layout itself to explain the basic sampling methodology.

You can also include some text to describe statistical manipulations behind the mapped data. Particularly if the map includes abstract concepts such as "walkability" (how well the built environment encourages its residents to walk), explaining how the categories were calculated is important. On map of walkability, the explanation might appear as something like "walkability was assessed from a statistical index that includes road density, land use analysis, and sidewalk survey audits." If you anticipate that terms like "road density" or "land use analysis" will not make sense to your audience, you can include even more detailed explanation.

I was involved with a mapping project to map obesity rates from hospital data across the Denver region by census tract (Anthamatten et al. 2020). The maps were intended to provide health practitioners and community members with a tool for evaluating the state of obesity in different parts of the city. In these maps, it was important to provide information about the data and methods for much of the audience. Our solution was to include a detailed caption directly below the map that contained a description of the map's contents:

> BMI, calculated from height and weight, is plotted on the CDC BMI-for-age growth chart to derive a percentile. Obese is defined as a BMI at the 95th percentile or higher. The percentage obese is calculated for each census tract by dividing the number of children and youth <21 years old with a BMI at the 95th percentile or higher by the total number of valid BMI measurements available.

This paragraph includes the key information about how the data were transformed to appear on the map. The readers of the map were given a clear idea about what the map showed and what the specific rates of obesity on the map meant.

You can include symbols and other map elements to communicate information about certainty on the map. Appropriate symbols can warn users to interpret the data with skepticism and caution. The *US Atlas of Mortality* map (Figure A.11) shows mortality from heart disease among white females in the United States between 1988 and 1992. The authors of the map were aware that some of the data were poor because there were relatively few patients in some parts of the country. Their estimates were based on the sample rates provided from women who died of heart disease; so, the fewer the cases in the Health Service Area, the more likely the results were to be affected by random chance and the less confident the authors were in their estimates. The authors added hatching—diagonal lines over the shaded portions of the map—to indicate where data were sparse. The symbol is explained at the bottom of the legend, and more detailed information is provided in other parts of the atlas, prompting readers to view rates in the areas with hatching with some caution.

Other types of symbology can communicate data uncertainty (Figure 10.4). Some geographic phenomena, such as the area where a religion is important or the range of a species of birds, are dynamic, ever-changing, or "fuzzy" to begin with. The borders around these areas rely on subjective decisions and can fluctuate over time. Traditional cartographic symbols—such as a line or areas—often imply that the regions

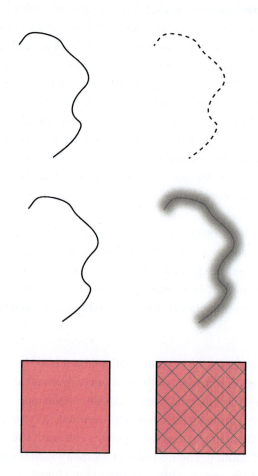

FIGURE 10.4
Graphical techniques for communicating uncertainty in maps.

are well-defined and permanent. A common practice is to demark the region with a dotted line. You can also intentionally blur the boundaries of an area to suggest that the boundary is fuzzy.

Summary

The rise of the Internet has enabled a virtual explosion in data. Around four billion people are connected to the Internet in some form, and the world produces an estimated 2.5 exabytes (an "exabyte" is equivalent to ten billion gigabytes) per minute (domo. com 2017), far more than people can ever reasonably use. One of the powers of cartography is that it can transform large amounts of data into a concise and meaningful form that enables the exploration of patterns and informing decisions. Appendix 2 contains a short list of some well-established repositories of geospatial data, which might help you get started on a cartography project. Some of the websites included offer only static map images, but the list contains several spatial data portals for use on geospatial and mapping projects.

Maps give power to data. With such great power comes great responsibility, of course. The proliferation of data has meant not only that it is much easier to collect data to produce maps but also that it is much easier to map poor quality data. Consequently, it is important that cartographers learn to be discerning collectors, users, and communicators of data.

Discussion questions

1. Identify some peers who have a similar interest in mapping some topic or phenomena. Take some time to identify potential data sources to contribute to your mapping project and discuss what you have found with your peers.
2. What are some of the ways that error and inaccuracy can be introduced into data throughout its trip from real-world data to your mapping project?
3. What are some additional tools you can use to communicate data certainty in mapping?
4. Have a look at Phil Germehl's map of histosols in Figure 10.3. You can also read the journal article in which he discusses how the map was ultimately misinterpreted and used (Gersmehl 1985). What suggestions would you make to the design of the map to avoid the problems that eventually resulted?
5. Examine a map that you produced or find a good example from a map in this book. What can you tell about the sources of data? Are you able to form an opinion about the data quality and what did you use to come to that opinion?

References

Anthamatten, P., E. Fiene, E. Kutchman, M. Mainar, L. Brink, R. Browning, and C.R. Nigg. 2014. "A Microgeographic Analysis of Physical Activity Behavior within Elementary School Grounds." *American Journal of Health Promotion* 28 (6):403–412. doi: 10.4278/ajhp.121116-QUAN-566.

Anthamatten, P., D. Thomas, D. Williford, J. Barrow, K. Bol, A. Davidson, S. Deakyne-Davis, E. McCormick, D. Tabano, and M. Daley. 2020. "Geospatial Monitoring of Body Mass Index: Use of Electronic Health Record Data Across Health Care Systems." *Public Health Reports* 135 (2):9. doi: 10.1177/0033354920904078.

domo.com. 2017. "Data Never Sleeps 5.0: How Much Data Is Generated Every Minute." https://www.domo.com/learn/data-never-sleeps-5?aid=ogsm072517_1&sf100871281=1.

Gersmehl, P.J. 1977. "Soil Taxonomy and Mapping." *Annals of the Association of American Geographers* 67 (3):419–428. doi: 10.1111/j.1467-8306.1977.tb01151.x.

Gersmehl, P.J. 1985. "The Data, the Reader, and the Innocent Bystander—A Parable for Map Users." *The Professional Geographer* 37 (3):329–334. doi: 10.1111/j.0033-0124.1985.00329.x.

Koch, T. 2016. *Cartographies of Disease: Maps, Mapping, and Medicine.* Redlands, CA: Esri Press.

Sanborn Insurance Company. 1904. *Sanborn Fire Insurance Map from Denver, Denver County, Colorado.* New York: Sanborn Map Company.

Sehdev, A.E.S., and G.M. Hutchins. 2001. "Problems with Proper Completion and Accuracy of the Cause-of-Death Statement." *Archives of Internal Medicine* 161 (2):277–284. doi: 10.1001/archinte.161.2.277.

US Department of Energy. 1979. "Peat Prospectus." Division of Fossil Fuels Processing. Washington, DC.

Veregin, H. 2005. "Data Quality Parameters." In *Geographical Information Systems: Principles, Techniques, Management and Applications,* edited by P. Longley, 177–189. Hoboken, NJ: Wiley.

Wood, D., W.L. Kaiser, and B. Abramms. 2006. *Seeing through Maps: Many Ways to See the World.* Oxford: New Internationalist.

11 | GIS and graphics software

Michael Goodchild wrote that the distinction between Geographic Information Systems (GIS) and cartography is now "hopelessly blurred" (Goodchild 2018). While the two fields were once viewed as distinct, where cartographers specialized in the design and production of mostly paper maps and GIS analysts focused on the management of and analysis with digital geospatial systems, developments in GIS ultimately catalyzed an era in mapmaking in which the two are viewed as a common enterprise.

Most people who work with GIS have a lot of practice explaining what GIS is to people outside of the field. Perhaps the most straightforward and simple definition is that it is "computerized mapping." A better, formal definition of a GIS is a functional one—a definition that describes what GIS can do:

> A GIS is a computer system capable of *capturing*, *storing*, *analyzing*, and *displaying* geographically referenced information; that is, data identified according to location. Practitioners also define a GIS as including the procedures, operating personnel, and spatial data that go into the system.
>
> *(United States Geological Survey 2018, italics mine)*

As cartographic output is one of the defining features of modern GIS, most geospatial analysts have some training in cartography. Performing spatial analysis can only get you so far if you are not able to effectively communicate the work with maps. Cartographers, on the other hand, must necessarily have significant training in GIS applications to tap into the bounteous geospatial data now available and the power that GIS offers for efficient maps production. There are multiple books on "GIS cartography" (an excellent example is Peterson 2014), and most university courses in cartography are deeply embedded within GIS programs. While there are still examples of successful work in cartography that relies on manual craftsmanship (see Harris 2006, for example), such work is more of an artisanal craft, now rarely practiced.

A lot of modern cartography is assembled, edited, and exported from geospatial software such as *ArcGIS,* or from its open-source counterpart, *QGIS*. While geospatial software has some graphical functionality, they are generally limited compared to applications dedicated to graphics. Graphical software can give you additional control over the maps and other graphical elements, enabling you to transform a workable GIS-based map into refined work. Working with graphics applications can also provide important foundational experiences, helpful as you learn a range of digital tools relevant to modern digital cartography. This chapter provides an introductory overview of GIS and graphical software, with a focus on terms and concepts relevant to their application to the modern practice of cartography.

An overview of GIS concepts and software

The aforementioned definition of GIS includes distinct functions: data capture, data storage, data analysis, and data display. Each of the GIS functions in that definition is important to the business of cartography. **Data capture** refers to the transformation of real-world information into a form that a GIS system can manipulate. Examples of data capture include digitizing paper maps, using **geographic positioning systems** (GPS), or data from **remote sensing**, such as from aerial photography or satellite imagery.

Data storage refers to the ability of GIS to keep spatial data in a form that can be retrieved and manipulated by a computer. How does one transform the real world, in all its boundless complexity, into an operational model that today's computers can understand? This basic problem echoes the underlying theme in cartography of thinking about how to model the world with abstract symbology.

Two dominant **spatial data models** have emerged and remain prominent. **Vector data** are built from lists of coordinate pairs. GIS software interprets single coordinate pairs as points, while lines and polygons are built from series of coordinate pairs. Points, lines, and polygons are collectively referred to as **feature geometry**. Readers familiar with graphical software may be aware that a similar model exists in graphical software ("vector-based graphics"), the subject of the next section. Figure 11.1 illustrates geographic data stored with a vector data model. Country boundaries are represented by a polygon, which is comprised of points, whose location is specified in the database by geographic coordinate pairs. The polygons are defined by points that are connected to form a boundary.

Raster data, by contrast, have continuous coverage of the mapped surface. In the raster data model, the entire mapped area is divided into a consistent grid, comprised of **pixels**, the cells of the grid. The raster data spatial model is comparable to raster-based graphics and displays (such as what you observe on a computer monitor). Each pixel contains a single attribute to represent the data. In the example provided in Figure 11.2, the map shows a raster data map of elevation of a portion of the Andes. The value in each cell represents elevation in meters above sea level. In this example, the resolution is one square kilometer.

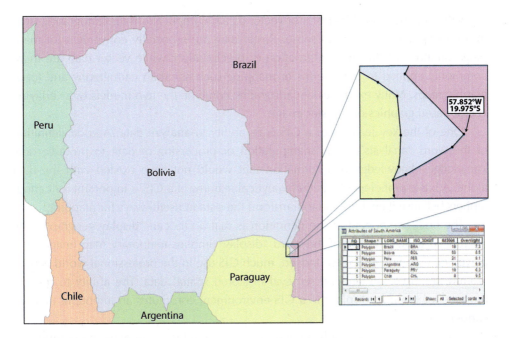

FIGURE 11.1
An illustration of country boundaries displayed in a vector GIS.

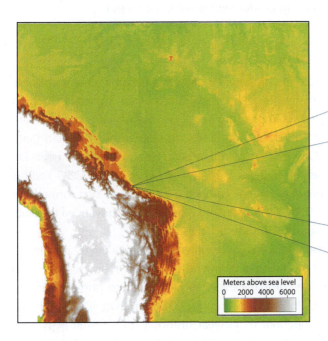

FIGURE 11.2
An illustration of a raster data GIS.

Whether you work with raster or vector data has implications for cartography, both in a conceptual and practical sense. Raster data are especially convenient for mapping continuous data such as elevation or precipitation, while vector data are more commonly applied to other forms of mapping such as urban cartography and topographic maps. Many examples of cartography combine the two models by overlaying vector-based graphics on a raster image.

One of the key features of a GIS is its ability to **analyze** data. A good definition of the word "analysis" is the manipulation or processing of data to produce new knowledge—knowledge or information that would not have existed with the data alone. A basic proficiency with the analytical features of a GIS is important for cartographers because it enables you to transform the data in useful ways for mapping.

The final component of the definition is that a GIS can "**display** geographically referenced data." The use of the word "display" implies temporary map images displayed on a computer screen. Indeed, much GIS use in the professional world is built around the rapid, on-demand retrieval of stored spatial data. The ability to quickly produce legible map displays in a GIS environment constitutes an important modern cartographic skill.

The functions of GIS applications apply to a broad range of activities pertinent to work in cartography. A GIS provides extensive and rapid access to data, enables cartographers to perform projections or recalculate thematic data in a way appropriate for the map, and, ultimately, facilitates in producing cartographic output.

GIS software applications

The potential importance and role of computers in mapping became clear with some of the early developments in computer technology. The original innovation behind modern GIS is usually attributed to Roger Tomlinson, an English cartographer who worked in Canada. During the 1960s, Dr. Tomlinson developed the Canadian Geographic Information System (CGIS) to assist with land and resource management. Dr. Tomlinson is often referred to as the "father of GIS" for his work on the CGIS and his contributions to the development of parallel computer applications that formed the basis of GIS software (Price 2019, 3). Tomlinson and other scholars advanced the state of geospatial software throughout the 1960s and 1970s.

An especially important moment in the development of GIS was the establishment of the ESRI by the then 21-year-old Jack Dangermond in 1969. Originally conceived as a land-use consulting firm, the company eventually became the primary GIS application used around the world. In 2017, ESRI controlled more than 40% of the GIS market share (Forbes 2018). Throughout its 50-year history, ESRI invested heavily in research and development. The company frequently publishes updates, plug-ins, and extensions to its software to keep up with the ever-changing GIS environment. Along with the main GIS desktop application, *ArcGIS*, their line of products now includes a variety of associated products and extensions to support GIS work with online services, 3D rendering, web mapping, and online data services ([ESRI] 2018). There are several notable GIS applications from other companies, often designed for a focused set of tasks or specific purposes, including *Global Mapper*, *GRASS GIS*, *MapInfo*, *GeoMedia*, and *Manifold System*, among others (GIS Geography 2018).

The most significant recent development in the GIS software is the emergence of open-source GIS, most notably *Quantum GIS* (QGIS). The project began in 2002 in tandem with other open-sourced geospatial data projects. The project is licensed under the GNU (which stands for "GNU's Not Unix!") license, an open-source license intended for use by the general public. QGIS has had a critical impact on the GIS user landscape and has ultimately made GIS accessible to many more organizations and individual users than in the past. QGIS can perform many of the functions of its proprietary counterparts and supports both vector and raster data models in numerous file formats.

An overview of digital graphics concepts and software

How much refinement you wish to give your map should be driven by the workload and the needs of the project. For day-to-day tasks that require you to quickly map a large variety of geospatial data, it is usually not worth the extra effort required to edit the map in a graphics application. If, however, you wish to include your map in a print publication, or perhaps the map will serve as a centerpiece to a widely distributed work (the map has your name on it, after all), then you might want to take the time and effort to bring additional refinement to the map. Having some skill and experience with graphical software applications, such as *Adobe Illustrator*, *Correl Draw*, *Sketchbook Pro*, or their open-source counterpart, *Inkscape*, can empower you with more complete graphical control over the map design.

Digital graphics have accompanied the development of computers since the 1950s. Within a decade of the first computer systems, vector graphics were developed to produce a low-memory means for storing graphical files. Encoding a set of critical points about the shapes of objects demands storing fewer bits of data than raster-based graphics, which (without any sort of compression algorithm) require storing information about every pixel displayed. An American computer scientist named Ivan Sutherland, who developed one of the first computer graphics software applications as part of his PhD thesis at MIT in 1963, is often referred to as a pioneer of computer graphics. In the following years, digital graphics was applied to many domains in computing. The financially lucrative video game industry provided significant market pressure that led to sophisticated graphics technology and production. Hundreds of graphics applications have emerged for a huge range of purposes, ranging from typography, video and film production, photography, web page production, to the production of scientific diagrams and maps. Like most of the technologies surrounding cartography, the world of digital graphics moves quickly—as new technologies continually emerge, old ones become less relevant and often fade into obsolescence.

Concepts in digital graphics

As is the case with just about any digital technology, knowledge about key terms and concepts can prove essential to learning to use software and keeping up with the ever-changing digital landscape. The smallest unit of programmable graphic space on a computer screen is called a **pixel**. Each pixel contains three tiny lights, called **diodes**, each of which emits a single primary color: red, blue, or green. Pixels are miniscule; the display resolution on computer screens refers to the number of pixels on the screen, usually given in terms of its width by height. A 1,920 × 1,080 computer monitor, therefore, contains over two million (1,920 × 1,080 = 2,073,600) pixels. Color is produced by altering the intensities of each of three red, green, and blue light diodes on a computer screen to produce the desired target color. Consequently, colors in digital graphics are normally given as RGB values, often in hexadecimal form.

Raster graphics programs—graphics that are based on a model of pixels or discreet cells, closely analogous to the concept of raster GIS—have become highly sophisticated and are capable of a wide range of graphical effects on images. Such effects include **texture mapping**, applying a pattern to a surface. Images can be enhanced through **sharpening** or **blurring** to produce desired effects. **Drop shadows** can be applied to graphical objects to give the appearance of a subtle shadow. Modern software includes sophisticated techniques for 3D rendering, manipulating the colors from patterns of **light effects** to produce a photorealistic image of objects.

While a lot of raster graphics functionality is useful for geospatial and mapping applications, several additional tools can be found in vector-based graphic applications. Vector graphics software can be used to perform basic smoothing and other generalization techniques on the map symbology, for example, and some software is particularly adept at manipulating text elements. When there is little need to involve complex spatial data, I have found that it can be worth doing all the work in a mapping project in a vector-based application. The final versions of all the original maps I have produced for this book were processed in a vector graphics program.

The fundamental structure of vector graphics is reminiscent of a vector GIS. While vector graphics do not include the ability to *geographically* reference objects,

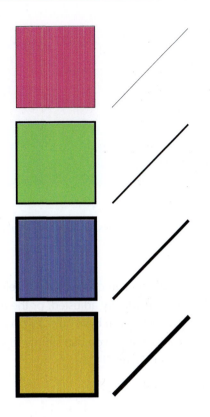

FIGURE 11.3
An example of simple objects with different fill colors.

they are built upon a Cartesian coordinate system based on points and lines which can be rescaled with little loss to graphical integrity. Individual objects can be constructed as points, lines, or areas. Each object can be assigned a **fill** color or pattern. The outline of each shape is called the **stroke**, which the user can manipulate independently from the fill color. In Figure 11.3, the boxes have different stroke widths. A variety of effects can be applied independently to both the stroke and fill of objects in vector-based graphics.

As in a GIS, the components of a vector graphic can be organized into layers, graphical elements ordered from top to bottom. Layers enable you to reorder the graphics to control their final appearance, facilitating techniques such as masking. They also facilitate the use of light effects and **transparency**, which can serve as an effective tool for reducing the visual weight of an object or text while revealing the graphics underneath it.

When you work with complex graphics with many components, the layers can play an important role in keeping track of and controlling the components of the work.

FIGURE 11.4
An example of layers panel in a vector graphics application (*Adobe Illustrator*).

Each layer can be named so that you can control specific elements, such as the text in a legend. Layer effects can be individually adjusted, or the entire layer can be toggled on or off, useful for adjusting a map. If you have built a map and want to alter the layout by removing a north arrow, for example, you can simply toggle the layer off as long as the arrow resides in its own layer (Figure 11.4).

Hexadecimal color codes

Working with graphics applications often requires extensive work with colors. As discussed in Chapter 7, the RGB color model was designed for display graphics. To make programming match the underlying architecture of computers, RGB values are commonly programmed and reported as a series of six digits in **hexadecimal**, and it is worth taking a small detour in the text to discuss the relevant logic behind it.

We are accustomed to using a *decimal* numerical system, based around the number ten. In the decimal system, we have specific symbols for the numbers 0, 1, 2, 3, 4, 5, 6, 7, 8, and 9. There is no unique symbol for the next number, ten, and so we represent the number with a combination of two number symbols, "10." Once the numbers advance to 99, we again run out of symbols, and so we add a digital to represent the next number, 100.

A commonly used numerical system is binary, a base two system. There are two symbols in binary, "0" and "1," which is a convenient system for computer programming, since computers are built upon bytes, which can either be "on" or "off." Counting through the first few numbers in binary is 0 (zero), 1 (one), 10 (two), 11 (three), 100 (four), 101 (five), 110 (six), 111 (seven), 1000 (eight), 1001 (nine), and so on.

Hexadecimal is a *base 16* system, which requires unique symbols for the first 16 numbers. The first ten numbers are the same as our decimal numbers, after which

TABLE 11.1 Examples of numbers in decimal and their conversions to hexadecimal

Decimal	Hexadecimal
1	1
2	2
...	
9	9
10	A
11	B
12	C
13	D
14	E
15	F
16	10
17	11
...	
99	63
100	64
...	
153	99
154	9A
155	9B
156	9C
157	9D
158	9E
159	9F
160	A0
161	A1
...	
254	FE
255	FF

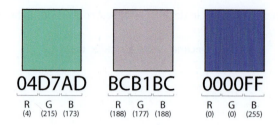

FIGURE 11.5
Examples of different hexadecimal codes and their colors.

they borrow from the alphabet: the number 10 is therefore represented as "A," 11 as "B," and so on, through 15, represented as "F." As with the other systems, the number advanced to the next digit when there are no other symbols, and so the hexadecimal number "10" represents the number 16.

In the hexadecimal system used in the RGB color model, 16,777,216 distinct values can be conveniently communicated with a six-digit hexadecimal code. The first two digits represent the red value, the middle two the green value, and the final two the blue value, each of which ranges in value from 0 to 256. The hexadecimal color code for "black" is therefore 000000, indicating that there is no light from any of the three diodes. The code for a mid-level gray is 888888, with all channels firing equally at about half capacity, and for white, with all three colors at full luminous intensity, is FFFFFF. Figure 7.8, from Chapter 7, shows the hexadecimal values for several color patches.

Fortunately, it is not necessary to memorize hexadecimal color codes, but it can be extremely helpful to understand the underlying logic of the system as you work with color in a digital environment. If you have a basic understanding of hexadecimal and the RGB color model, you can make a reasonable guess of the color from reading the code. Try breaking down the hexadecimal code 04D7AD, shown in Figure 11.5. The first two digits, representing the red channel, equal 4, and so we can determine that there is almost no red in the color. The next two digits are D7, which is equal to 215, which means that the green channel is at a high intensity. The final two digits are AD, equal to 173, a mid-to-high intensity for blue. The color is a moderately intense bluish green.

There are good online resources to help you translate html color codes (see, for example, www.rapidtables.com/convert/color/hex-to-rgb.html). Most graphic applications have a dialog box for selecting and sampling color that enable you select or adjust color with a clickable interface or by entering the values for any of the available color models.

Raster graphics file formats

Static digital maps can be stored either as a raster-based image or a vector-based image. The highest-quality image raster files can be stored as a **TIF** or **TIFF** ("Tagged Image file") file. TIF files contain a lot of information about the graphics and are **lossless**, which means that compression should not cause any loss of graphic quality. Due to the high-quality graphics that they enable, publishers often request TIF files for images that will be printed. Another advantage is that they are **extensible**, which means that it is possible to extend the file by adding additional information or functionality to it. One important extension for mapping is the **GeoTIFF** format, which contains the graphical information of a TIFF with the addition of coordinate information. Maps that are released as GeoTIFF images, such as the images of topographic quadrangle maps provided by the USGS, can be opened in a geospatial application with the coordinate system in place, enabling them to immediately be correctly placed in the map space.

The main drawback of TIF files is the file size. Even modestly sized images can consume a lot of space. File size is seldom a problem for *storing* data, since storage is no longer a significant constraint in most modern working environments, but it can produce problems when the map or image is intended to appear on a web service and must be retrieved through a potentially slow Internet connection, such as on a web page or mobile phone. A common alternative raster image format that takes less storage space is a **JPG** or **JPEG** (normally pronounced as "jay peg"), which stands for Joint Photographic Experts Group. JPG files use a compression algorithm that preserves the quality of the image while reducing the file size. In many applications, you can adjust the compression rate of JPG files to build smaller files, but reducing file size comes with the price of some loss to quality.

There are a few other raster graphics file formats, but most mapping uses one of the file formats discussed earlier. Other formats include **GIF** (variously pronounced with a soft and hard "g"), which refers to "graphical interchange format." GIF files generally are constrained to eight-bit color, which means they can only support up to 256 different colors. They consume very little file space and are primarily suitable for simple diagrams. **BMP** (Microsoft bitmap) files were developed by Microsoft for use with Windows applications; they are easy to construct and can be saved in compressed format, but they generally consume a lot of disk space and do not perform well when rescaled. **PNG** ("portable graphics format") is another format developed for various graphics applications, originally designed for use on the Internet. PNG files can support transparency and a variety of other graphic effects.

Vector graphics file formats

Vector file formats generally produce higher-quality prints due to their scalable resolution. A basic vector graphics file format is the **encapsulated post-script (.eps)** format. EPS files were developed in the 1980s and have enjoyed widespread applicability and usability in graphics since that time. Key benefits of EPS files include their relatively simple encoding format and their breadth of use.

Another dedicated graphics file format is the **scalable vector graphics (.svg)**. The file format emerged in 1999 from an effort directed by the World Wide Web Consortium, a key group responsible for directing standards for the Internet. SVG files are text-based, open-standard files that use extensible markup (XML) language, specifically designed to function in coordination with other web-based standards and file protocols. Its architecture enables the files to be searched, indexed, and compressed. Due to these goals behind its development, its format has become broadly applicable and is now supported by all major web browsers.

Many maps are published as **portable document format (.pdf)** files. PDF files were initially developed by the graphical design software company, *Adobe*, for internal use, with the idea that the company could avoid working with printed documents by using electronic documents. The concept of the file was to provide a stable, low-memory, flexible, and universally readable format that could enable files to convey a variety of types of printed media such as text and images. The success of the effort led to the Adobe release of *Adobe Acrobat* in 1993, a program that enables users to convert their work into PDF form. Since then, the format has been updated several times, and now serves as a common standard for storing text and image documents. PDF files can contain any combination of text, images, and vector-based graphics. Because of their size flexibility, the files eventually became an important format for storing and sharing maps. **Geospatial PDFs** now enable users to find and mark locations, and make measurements on the map with user-specified coordinate system and locations (Adobe 2020). Many high-quality maps are now available as geospatial PDF files, including the topographic quadrangle series published by the USGS.

Several other vector-based graphics file formats were developed by software companies to function with their suite of products. Their portability to other applications varies. Examples include Adobe Illustrator (AI), Windows metafile (WMF), and Correl Draw (CDR) files. The file formats discussed here can be used with most open-source graphics applications.

Summary

The fundamental technical skill for cartographers is in geospatial science and GIS, but cartographers should be prepared to learn about and engage with a broad range of computer software and hardware. Because cartography requires producing graphical images, experience with graphical software applications is a good next step.

The expansion of cartography into web-based and mobile mapping means that cartography jobs often require skill in programming languages such as C#, Python, Javascript, and Visual Basic. Because of the importance of spatial data, data handling, and database management in mapping, experience working with databases and working in database languages such as Oracle, Structured Query Language (SQL), or Microsoft Access is often a plus. Producing 3D or animated cartography can require working with 3D graphics and video production software as well.

While few cartographers can claim expertise across the entire suite of applications relevant to mapping, performing the nuts-and-bolts work in cartography certainly requires a fundamental willingness to engage with a variety of computer applications and the ability to grapple with unfamiliar software environments as the need arises. The specific type of computer and technical skills you gain as a cartographer is likely to be driven by the nature of the job. Perhaps the most important strategy for learning the computer applications that enable map production is hands-on experience in building them. This book is accompanied by an online tutorial to guide you through the process of going from an idea for a map to a final map product using common GIS and graphics applications.

Discussion questions

1. The text has reviewed a variety of different types of skills and applications relevant to cartography. What kinds of mapping do you believe could benefit from sets of these technologies? For example, for what fields of mapping do you think geospatial applications are particularly important? What fields do you think rely heavily on graphics applications or mobile technologies?
2. What do you think are some additional advantages or disadvantages of working with cartography using vector graphics and GIS, compared to versions of the software that use a raster model?
3. Many of the applied components of cartography require a huge commitment to continually learning new software applications and keeping up-to-date with advances in the technology. What strategies can you adopt to keep abreast all the changes as a professional cartographer?

References

Adobe. "About Geospatial PDFs." Accessed May 28, 2020. https://helpx.adobe.com/acrobat/using/geospatial-pdfs.html.

[ESRI], Environmental Systems Research Institute. 2018. "ESRI." Accessed August 4, 2018. https://www.esri.com/en-us/home

Forbes. 2018. "#527 Jack & Laura Dangermond." Accessed August 4, 2018. https://www.forbes.com/profile/jack-laura-dangermond/#765e797205f8.

GIS Geography. 2018. "Mapping Out the GIS Software Landscape." Accessed August 4, 2018. https://gisgeography.com/mapping-out-gis-software-landscape/

Goodchild, M.F. 2018. "Cartography: GIS and Cartography (Relationship Between)." Accessed August 7, 2018. http://www.geog.ucsb.edu/~good/papers/474B.pdf.

Harris, A. 2006. "Ski Trail Map Painter." *Fortune* 153 (4):39–39.

Peterson, G.N. 2014. *GIS Cartography: A Guide to Effective Map Design, Second Edition*. Boca Raton, FL: CRC Press.

Price, M. 2019. *Mastering ArcGIS*. 8 edn. New York: McGraw Hill Education.

United States Geological Survey. 2018. "What Is a Geographic Information System (GIS)?" Accessed July 16. https://www.usgs.gov/faqs/what-a-geographic-information-system-gis.

12 | Examples from the field

One of the foundational skills for work as a cartography professional is resilience to deal with novel challenges, the creativity to sort through processes that do not work the way you expect them to (including the ability to work your way around software bugs) and, often, the patience for interacting with clients who have no idea what the work entails. Many projects may require you to break into a completely novel software application with which you have little or no experience.

In this final chapter, I focus on the non-technical components of the process of making a map, from inception to the final product, drawing from examples taken from my own experience and work. In this discussion, I highlight some of the considerations, design decisions, and challenges that comprise nearly any mapping project. The intent of guiding you through these stories is to give you "real-life feel" for decision-making behind making a map and to prompt you to think about the application of the concepts covered in this book. I recommend that you review Appendix 3, a set of criteria for cartographic review, in conjunction with the discussions that follow.

A map of historical wildfires in California

Several years ago, a colleague of mine asked me to help him produce some maps about wildfires in California for a series of research projects. The projects explored the politics, policies, and social pressures surrounding the expansion of suburban and exurban neighborhoods into the fire-prone hills of California, which ultimately resulted in exposing residents to forest fires (Simon 2016). The topic is complex and rich, invoking many facets of the study of geography. Dr. Simon wished to include a series of maps at different scales to show the geographical context of the region, voting behavior, and ethnicity.

The idea behind one of the maps, central to the goals of the project, was to clearly show the historical extents of wildfires outside of Oakland, California. As emphasized throughout this book, an important first step in any mapping project is to develop guidelines for the fundamental facets of the maps: the maps' purpose, audience, and cartographic medium. In the case of my colleague's fire map, all three were clear. The map was intended to provide an overview of the history of fires throughout the twentieth century around the city of Oakland, to be included in a book that explored the political ecology around occurrence of fires. The map should give special attention to the 1991 Tunnel Fire in Oakland, around which much of the narrative of the book was focused. The expected audience consisted of scholars and others with a keen interest in the topic, and so we could anticipate that the audience would be well-educated, have strong knowledge on the topic of wildfires, and generally possess good familiarity with the region. The map was intended to appear in print, but we anticipated that it might also ultimately appear online as a static digital map.

After clarifying these questions in our initial meeting, we discussed what map features should be included. The map would appear at a point in the text where the study area had been well illustrated with other maps showing the context. Consequently, the map only needed to include minimal orienting spatial information. Reducing the amount of contextual information—such as roads, borders, bodies of water, and labels—would serve to keep the reader focused on the main idea of the map: historical patterns in wildfires.

It is a good idea to definitively finalize the size and extent of the map in communication with everyone involved in a mapping project, as one of the first steps. As you begin to assemble data, particularly once you start working in a graphics application such as *Illustrator*, changing the extent of the map can result in lots of additional work. If the team decides to add area outside of a map project, for example, you may have to seek out additional spatial data, rework the projection and clipping work, and ultimately restart the process of moving the map data from a GIS to a graphics application. All the members of the team may not understand this, and so it is important to be clear about the importance of the decision about the extent early on. At our first project meeting, we examined Google Maps on a computer to gain a sense of the area of the fires he wanted to map to define the boundaries of the map.

With good idea of the nature of the project and some of the source data in the form of printed maps in hand, we sought digital spatial data on some common types of orienting features, including political boundaries, bodies of water, urban areas, elevation, and major roads. We downloaded political boundary and road data from public and open data sources, such as the United States National Atlas. Among the files we prepared was a georeferenced image from Google Maps, intended to provide us with a reference for street names and to enable us to digitize other features.

The heart of the map would show the extent of fires in the region. Dr. Simon provided us with a series of original documents from the US Department of the Interior and the National Park Service showing fire extent, as well as some work that he and others had put together from other sources. Because these documents did not exist in digital form, we scanned them into digital images, georeferenced them, and then manually digitized the extents of the fires. We wanted to include some bodies of water to provide further context to the region, and so we digitized the larger lakes in the region, leaving the smallest lakes out of the map. Figure 12.1 shows a GIS data view of some of the initial work. The Google Maps background was taken from a screenshot and

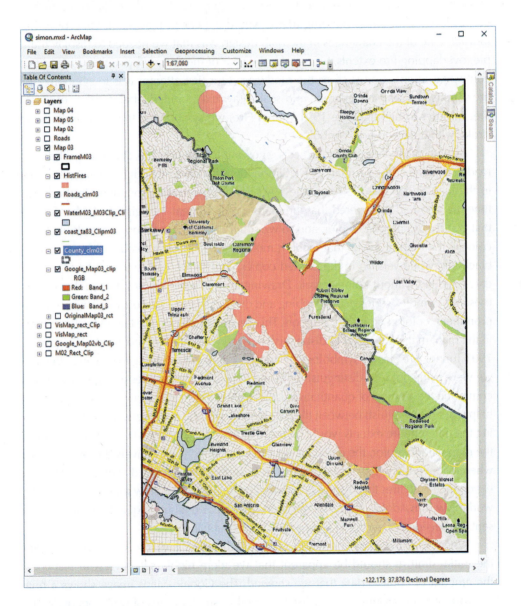

FIGURE 12.1
A screenshot of the GIS data view with several layers of data for the Flame and Fortune Mapping project.

then georeferenced, from which lakes were selectively digitized. The fire extents were digitized from paper sources (appearing as red blobs in the middle of the image), and road and county boundary data were assembled from publicly available data from the United States National Atlas.

The next step was to move the GIS work to *Adobe Illustrator* to produce a basic mockup of the map to present to Dr. Simon (Figure 12.2). We produced the map in grayscale with the understanding that it would appear in black-and-white print in a journal. One of the challenges in the map design was that we needed to display multiple, overlapping fire extents, along with the year of the fire, without obscuring critical contextual information such as the roadways. Our first attempt was to symbolize the fire extents in a semi-transparent, mid-scale gray that included labels for the year of the fire. In order to highlight the 1991 Tunnel Fire, we symbolized the fire using a wider stroke with a dashed line symbol.

As with most cartography projects, showing the map to the client and other map readers yielded critical guidance. Dr. Simon identified several additional fires to include in the map and wanted to make the 1991 Tunnel Fire more prominent. In the meantime, he also had discovered that the maps would appear in digital and other formats that could accommodate color, and so we sought a means to leverage the ability to use color in the map in a way that made it more legible. The map would also appear in a black-and-white version in the printed copies, and so we had to design to accommodate legibility without color as well. Drawing from other examples of map designs that present a complex series of overlapping and distinct shapes, we converted the areas into prominent, colored boundary symbols. This version improved the legibility and contrast in the map, making it easier to distinguish between the different fire events. We gave the 1991 Tunnel Fire clearer symbology by applying dark-gray shading to that area. Because we were using color to distinguish between different types of fires (our goal was not to communicate any sort of quantitative data), we were able to effectively use color hue to help readers distinguish between the different fires for the color versions. While perceiving the different fire extents would be more difficult for readers who viewed versions of the map without color, the date labels for the fires effectively conveyed its key message when it converted in grayscale: that the 1991 Tunnel Fire occurred after a decades-long experience with fires throughout the immediate region. After a few rounds of refinement working within *Adobe Illustrator*, the team was satisfied with the quality of the map and sent the version appearing in Figure 12.3 to the publisher.

When you juxtapose the final version of the map with the GIS screenshot and the initial draft, it should be apparent that we were highly selective with respect to the amount of context data that we showed. The Google Maps image shows a huge network of roads around the city itself, as well as small bodies of water peppering the region. We only included the major roadways and the largest bodies of water, leaving most of them out to avoid cluttering the map with too much extraneous detail.

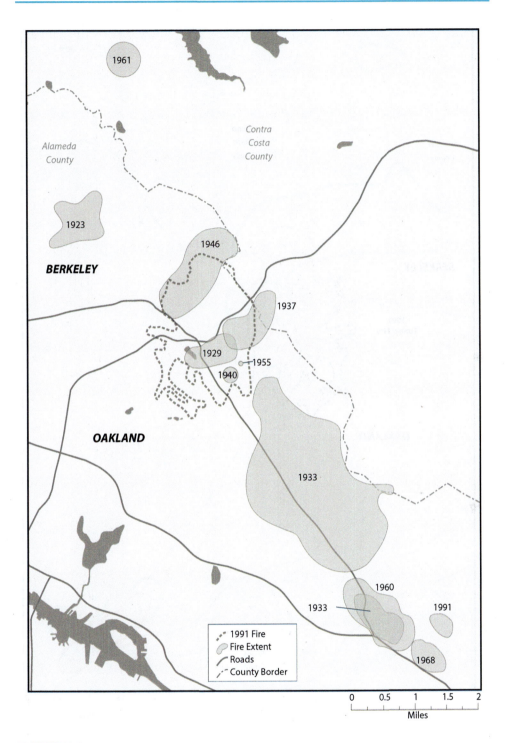

1961

Contra
Costa
County

Alameda
County

1923

1946

BERKELEY

1937

1929

1955

1940

OAKLAND

1933

1960

1933

1991

1968

1991 Fire
Fire Extent
Roads
County Border

0   0.5   1   1.5   2

Miles

FIGURE 12.2
An early draft of one of the *Flame and Fortune* maps.

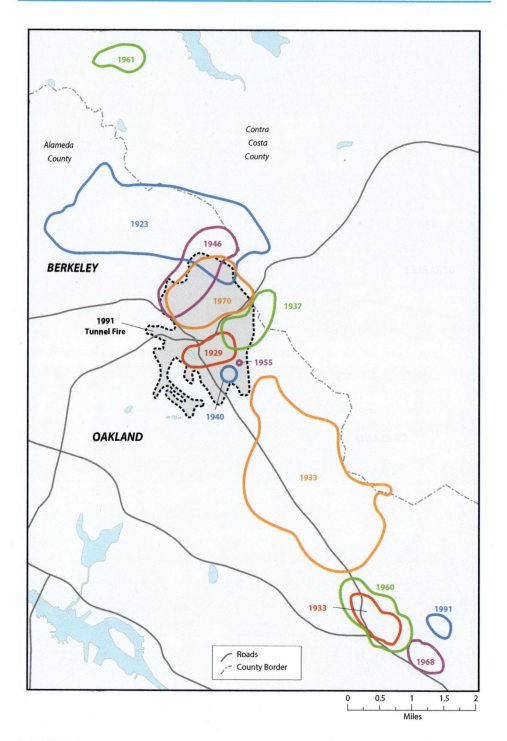

FIGURE 12.3
The final draft of the map "Periodic Fires in the East Bay Hills."

We modified the text and the size of the symbols so that the visual hierarchy would guide the readers to the fires, by giving the county labels, the legend, and other features a lighter color that made them blend into the background. We also removed a couple of legend items to simplify reading the map. The narrative of the text focuses around the roles that these cities, their policies, and the politics played in the occurrence of wildfires, and so it made sense to give the labels for the cities of Berkeley and Oakland more visual weight than the other contextual map features.

A map of beekeeping activity across Denver

A few years ago, I began to collaborate with a palynologist colleague of mine who became interested in studying bees through her interest in pollen; she studied bees to understand foraging patterns and resource availability for bees in urban settings. The study of urban bees and beekeeping is deeply linked to the environment. The success of beekeeping project and the health of bees are driven by factors such as the availability of pollen, nectar, and water, as well as exposure to pesticides and parasites (particularly the Varroa mite, which is a significant problem for bees in North America). Since beekeepers are people, bee health is also driven by social factors, such as access to good information and resources for beekeeping. One of the goals of the project was to produce maps of Denver showing where beekeeping occurs, as well as whether there are systematic patterns in problems (such as hive death or *Varroa* mite infestations), and to examine whether there are any observable links between environmental factors, beekeeping success, and bee health.

My colleague and I worked with the local beekeeping community to collect around 200 surveys from beekeepers in and around the city of Denver. The survey asked respondents for basic demographic and social data as well as observations about their beehives. In order to produce some basic maps of beehives across the city, we needed to have information about *where* the beekeepers kept their hives. Asking people where they keep their hives is a little tricky; theft of beekeeping equipment (and even the bees themselves) is a nontrivial concern for beekeepers, and so we had to frame the question to yield identifiable geographic features that we could map with a reasonable degree of accuracy, but which did not expose the survey-taker to risk of theft.

We asked respondents to "specify the street intersection (with the city or town) nearest to your bee hive(s)." While data on the "nearest street interaction" would not provide very accurate locations, data on nearby street intersections gave us enough information to geocode and map points from the surveys. A benefit of this type of question was that respondents were more likely to respond since a nearby street intersection would not give away the precise location of their hive. Furthermore, most people in the USA are very familiar with the street networks and would be likely to be able to provide an accurate response. Once we had collected the survey data, a student researcher utilized Python's integrated development environment (IDE) to transform street intersections into latitude and longitude points. Automated geocoding normally does not produce complete results for several reasons (the street spellings were incorrectly recorded in some of the survey responses, for example), but we were able to find the unreferenced addresses in Google Maps, from which we were able to map all of the responses.

Our mapping task at this stage in the project was to produce some maps to see if we could find patterns in beekeeping. The purpose of the maps would be to facilitate our next steps in the project, such as what facet of the data to analyze. Our goal, therefore, was more of an *exploratory* than a communicative one; we needed to gain a basic sense of what the data were telling us to guide us to the next steps in the project. The audience was our small research team of faculty and students involved with the project, and we could produce all the maps on computer screens for small meetings. For an exploratory task such as this, the mapping needed to be quick, flexible, and private. We needed to get the data into a form that we could easily transform if we wanted to examine a facet of our own data or juxtapose the data with other environmental data, such as green spaces or social data. Because we did not plan to use the maps to communicate our work to others at this stage, we did not have to spend time refining the map or clarifying its nuances; the team was very familiar with the data and the region we were mapping.

In preparation of our team meeting, we collected some basic reference information such as county boundaries and major roadways. The initial task was to gain an *overall* sense of the patterns of beekeeping across the city, and so we performed a basic smoothing algorithm in *ArcGIS* to generalize the points. The result was a raster map that symbolized the areas with a denser collection of beehives in a darker shade and the areas with fewer beehives in a lighter shade. The resulting map effectively achieved what we needed (Figure 12.4). The dark lines represent county boundaries, the lighter lines represent major roadways, and the red points represent reported locations of beekeeping. As long as we enabled the team to visualize broad patterns across the area, we did not need to include many of the traditional map elements or dedicate a lot of time effort to the design. Many traditional mapping elements (labels, a title, a scale bar) were omitted.

Armed with a strong knowledge of the geography of the area, some key ideas were immediately apparent to the team. For example, there appeared to be very little urban beekeeping along the western and northern edges of the city of Denver (the polygon in the middle of the map shows the city boundaries), where there are many low-income neighborhoods. Additionally, there was quite a lot of activity just south of the city in the nearby suburbs, perhaps due the locations of beekeeping businesses and an active beekeeping community. This kind of exploratory mapping enabled us to generate these insights quickly and efficiently, prompting us to consider both research questions (for

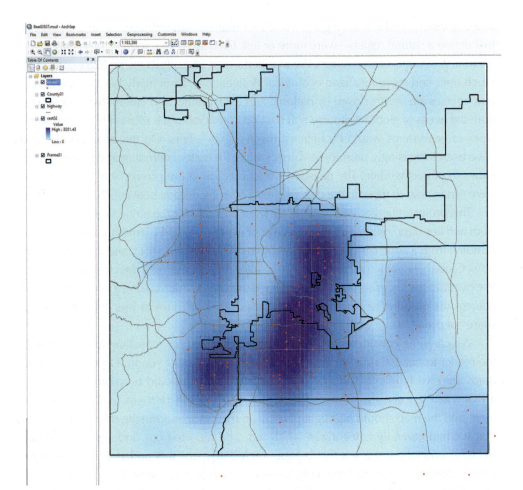

FIGURE 12.4
Exploratory map work in a GIS for the Denver Urban Beekeeping project.

example, what role does poverty play in beekeeping?) as well as some guidance about where beekeeping efforts would best serve the needs of the city. Having the GIS on hand with the geocoded survey data also gave the team efficient access to maps of other facets of the data, such as whether there are systematic patterns in the rates of hive death or observed infestations with Varroa mites.

A map of protected areas across the globe

The next map project was a collaboration with a colleague to examine biases in the distribution of protected areas worldwide (Anthamatten and Hazen 2015). We were interested in the question of which among the world's 14 biomes—assemblages of flora and fauna that define a natural region—were more or less likely to receive formal "protection" status from governments than others. We also performed an analysis on the 825 ecoregions, much finer and more localized ecological designations. Particularly because conservation policies are mostly driven by national and local politics and culture (which often ignore the needs of global discourse and data on conservation), it seemed likely that some types of ecosystems would be better protected than others. On the other hand, there have been increasingly prominent voices calling for better international coordination in efforts to protect wildlife. The question we focused on in this project was whether there were observable changes in the degree of bias over the decade between 2003 and 2012.

This project relied on secondary data from GIS data for the entire world that pulled from two key data sources: a GIS data layer of ecoregions and biomes that had been assembled from a large collaboration of scholars (Olson et al. 2001), and the United Nations Global Database of Protected Areas (World Database on Protected Areas 2012). We overlaid the layers in GIS and performed area calculations using a simple model to address the question of whether some biomes received more or less than "the global share" of protected areas.

We carefully considered what types of maps would be interesting and relevant to the reader to better tell the story of the project in a scientific report. On the one hand, maps are clearly relevant to the topic at hand; readers should be prompted to think about where conservation occurs, how it changed over the ten-year period we examined, and the map should highlight patterns in the changes. One the other hand, given that we had carefully categorized, analyzed, and reported our findings in the paper in charts and tables, we wanted to avoid including maps that added little meaningful information to the report. Especially in venues such as a scientific journal article, it is important to carefully consider whether the map contributes to the story at hand.

We decided to include two world maps. One map showed the current protected area status of ecoregions in 2012, distinguishing which ecoregions had reached an international goal of 17% of protected area from the ones that had not, and which ones had no protection at all. We also wanted to include a second map that displayed (1) new protected areas added during the study period (2003–2012), (2) areas that were protected in 2002 but which lost their protection status during that period, (3) areas were consistently protected both in 2002 and 2013, and (4) areas not protected at all during the time period. We had produced the data for this map in our analysis, since the point of the paper was to examine systematic biases in losses and gains. The second map was far more complex than the first due to its small scale and the detail of the input data; it was built from overlays of detailed protected area data collected from national governments overlaid with biomes. In a page-sized map of the entire globe, our hope was that some patterns would emerge from this huge amount of relatively high precision spatial data.

Because the map covered the entire globe, the projection choice was important. Since the map was about showing protected areas on the ground and the aerial extent was an important part of that, we chose an equal area projection, the Mollweide. The main distortion in a Mollweide projection is near the poles, where shapes are increasingly distorted. This was okay for the project, because the key subject area of the map was in latitudes away from the poles. Because the projection mimics the spherical shape of the globe, the distortion could also serve to remind the viewer that they are viewing the spherical earth.

We learned that the map would appear in color in its online version. It was placed in the middle of a scientific report on protected areas that would be read by environmental scientists, geographers, and policymakers interested in global-scale protection efforts. Consequently, we counted on the readers having a strong familiarity of the topic and with the earth and its features. We did not include any labels on the map at all and omitted a scale bar altogether, but did include a descriptive caption, which read:

> Map of protected area gains and losses between 2003 and 2012. The green areas show protected area that was added to the World Database on Protected Areas (WDPA) between 2003 and 2012. The red area shows protected area that was removed. The dark gray area shows protected area that was in the database in both 2003 and 2012.
>
> *Source: Data from Olson et al. (2001) and WDPA (2012)*

Because the text was described in the caption, we also did not include a title directly on the map itself. We included only a very basic and faintly symbolized graticule, spaced at intervals of 30 degrees, to provide some guidance to the shape distortions and to improve figure-ground.

Color decisions were important in this map. We wanted the reader to focus attention on the protected areas that were lost or gained, and so we gave the areas that were not protected shades of gray. Areas that were not protected at all during the period were light-gray, and protected areas with no change during the study period were symbolized with darker gray. The critical goal of this map was to highlight areas where there was *change*. Areas that lost protected status were symbolized with bright red, and protected areas that were added were colored bright green. A benefit of this color choice was that even very tiny areas (perhaps as small as 10 or 20 square kilometers) would be visible on the map. The map was really about the changes occurring on terrestrial land, and so the ocean was de-emphasized with a light, desaturated blue color. The final version of map appears in the map gallery as Figure A.12.

Due to the limitations of how the graphics were symbolized, some parts of the map appear to have exaggerated representation. A modest amount of conserved area ultimately appears much larger on the map (Central Europe, for example, appears to be almost entirely covered by protected area on this map), resulting from a type of "graphical processing error" on the map. Conveniently, this worked in our favor to produce a type of generalization we needed: exaggeration. I believe that this map ultimately achieved its goal because the data served to bring out patterns in conservation change. The map shows that conservation in Europe is stable and fragmented; large areas are covered with a broad smattering of gray. The greatest additions to protected areas were in Asia and around the Amazon in South America, where large patches of green are visible. North America, on the other hand, appears to have lost vast swaths of protected area.

As I witnessed the reactions of my colleagues and others to this map, mostly from the USA, I began to notice that readers nearly always expressed shock and horror at what appeared to be occurring in North America, with much of the western part of the US riddled with an ominous-looking bright red, appearing as if the US was abandoning conservation on a massive scale.

In actuality, the management of protected areas in the Western US changed very little during the time reported on the map. The data on the map were originally derived from national governments who were responsible for categorizing the protected status of land according to international standards. In the mid-2000s, policymakers and others in the United States determined that much of the area previously considered "protected" did not meet international standards, and so they removed the protected designation in 2009 (Coad et al. 2009), which then made its way into the international database on protected areas used in our research project. The apparent loss to protected areas appearing on this map (and ultimately in our analysis) was not a reflection of any sort of meaningful change on the ground, but rather represented an effort to improve to the rigor of data reporting. This highlights one responsibility of modern cartographers and map-readers that draws from Gersmehl's ideas articulated earlier in the text: it is important to always bear in mind the context in which the map could be viewed and to guide the reader from cases of potential or likely misinterpretation of the fundamental ideas. In retrospect, I should have included some text or a disclaimer on the map to guide readers about this, in anticipation of the possibility that not all the map readers will have read the entire article or considered the map in its intended context.

An application of modern GIS and cartography to John Snow's classic cholera map

Geographers, public health scholars, and cartographers frequently cite John Snow's map of cholera deaths in London as the first well-known example of "analytical mapping," mapping intended to analyze patterns in some phenomenon with the intent of gaining novel insight to the topic being studied (a reproduction of the original map is provided in Figure 1.4). In eighteenth-century London, John Snow was able to use mapping to make the case that there was a link between cholera deaths and the Broad Street pump in an effort to convince his contemporaries that cholera was linked to exposure to contaminated water supply (see Johnson 2006). His work is often lauded as a prime early example of the power of evidence-based, analytical mapmaking. Viewing data on cholera deaths in terms of their *locations* completely changed the story of the data. The map clearly shows that the deaths were clustered around that pump, situated in the center of the map.

I was once invited to contribute a chapter that focused on the role of geography in the study of health, to be included in a book for public health students (Anthamatten 2015). Each chapter of the book introduced a different disciplinary focus in public health (such as epidemiology, economics, and psychology), and my charge was to introduce the readers to the ways in which geography, and specifically GIS and mapping, contribute to the field. The readers would consist primarily of European university students preparing for a public health career, and so I could assume that many of its readers would have little or no background knowledge or experience in cartography (or even that they understood the acronym "GIS"). The other authors referred to Snow's map at multiple points in the other chapters of the book, and so I decided to include an updated version of the map to demonstrate the power GIS analysis and illustrate how modern cartographic techniques could used to present the story in an even more powerful way than the original.

I georeferenced a JPG image of Snow's map in a GIS, and then digitized the pump locations, the street network, and cholera deaths. Because I was able to relate a few key points on the original map to modern locations in London (including precise data on the location of a memorial at the site of the original Broad Street pump!), it was not too difficult to georeference the image and calculate the scale of the map. I used the GIS to build **Thiessen polygons** around the pump locations; Theissen polygons provide a straightforward method for delineating areas that are closest to each of a set of points, providing a crude model for the areas where residents were likely to have sought water from the pump. With a Thiessen polygon feature class on hand, it was a straightforward task to calculate the number of cholera deaths within each polygon in the GIS.

While Snow's analytical map is viewed as a classic map that represented a revolutionary idea at the time, there is room to improve the design, drawing from modern GIS technology and the principles of cartographic design discussed in this book. The main purpose of Snow's map was to explore the way in which cholera deaths were related in space to the Broad Street pump, suspected to be the source of the infection. In the original black-and-white map, the pumps were symbolized as black squares, and the deaths were shown as tally marks—like an in-map bar graph. The stacked line symbols representing cholera deaths are distinguishable from the other features of the map, but it is difficult to distinguish those symbols from the pumps, designated as a small circle with a label. The map suffers from poor contrast, legibility, and visual hierarchy. The streets are meticulously drawn and labeled, but the inclusion of the information has a lot of visual weight that draws attention from the main subject of the map: the cholera deaths and pump locations.

In this Snow "reboot" map, I symbolized the streets as white space, and the areas in between the streets (densely packed with buildings at the time) are shown as a light gray color, serving to provide information on the street network while consuming little visual weight (see Figure 12.5). I transformed the roughly 530 deaths recorded on John

FIGURE 12.5
A modern revision of John Snow's map of cholera in eighteenth-century London.

Snow's map from strike marks to small black points, easily distinguishable from the background. The pumps are also symbolized as points, but with a much larger symbol, in the shape of a star and colored bright red. The red water pump symbols are the only instance of color on the map, ensuring the pumps are easy to identify, sitting high in the visual hierarchy. To improve the analytical legibility of the map, I symbolized the Thiessen polygons as dashed lines and placed them over the entire map. Finally, I placed numbers reporting the total number of deaths within each Thiessen polygon directly on the map. This technique enables the reader to immediately see that the majority of all the deaths in the study area occurred in the catchment area of the Broad Street pump. Finally, a legend was included to help the reader decipher the symbols, as well as graphical scale.

The point of this exercise was certainly not to emphasize fault in the cartographic work of John Snow, whose meticulous data collection served as a model for generations of epidemiologists and public health scholars, but to demonstrate the power of maps to tell a story about data in a clear, compelling, and honest way. With a little careful thought about basic cartographic decisions, such as what features to include or omit, how to symbolize the features, and what additional information is needed to help the readers understand the map, his story becomes clearer and even more compelling.

Maps of children's physical activity in urban schoolyards

The final mapping example I will discuss came from a collaboration with members of the School of Landscape Architecture in my university, which had engaged in a years-long project to renovate the schoolyards throughout the city of Denver. Along with the schoolyard construction, a research group was studying the health impacts of the renovations. Part of the work involved careful data collection of children's physical activity on schoolyards in carefully designated "observation zones."

One of the questions that came out of the project was how to determine whether there was a link between the patterns in construction and children's physical activity. We hoped our work could serve as guidance for landscape architects who wished to design schoolyards that encourage children to engage in physical activity. We decided to map the activity in the schoolyards along with the spatial facets of the schoolyard environment that included constructed equipment, the size of the area, different surface types, and the presence of trees and planted material.

The purpose of the maps was both exploratory (to examine patterns in physical activity and schoolyard design) and communicative (to demonstrate the value of schoolyards and good schoolyard design to other researchers and policymakers). The maps needed to be clearly legible and accessible to broad audience, and effectively convey a lot of data (information about physical activity) and spatial context (the design and structure of the schoolyard).

We began by painstakingly digitizing maps of all the schoolyards in the study, along with the design plans provided by the architects involved with the project (architects commonly work in design software called *AutoCAD*, and they provided us with files from their work). We were also provided with high-quality aerial photographs of the schoolyards. Working with this material, we built the maps in a GIS, with which we combined the observation data we had collected in children's physical activity.

Producing clear, honest maps of the schoolyards was a difficult and required some care. The data were collected by researchers who spent time on the grounds recording how many children were in each zone, and how many among those children were engaged in physical activity. Because the observation zones were different sizes, the schoolyards had different population sizes, and the observations themselves came at different times, we had to come up with a comparable measure for physical activity in the zones. We devised a *relative* measure that showed whether children were more or less likely to practice physical activity in each observation zone, compared to the overall physical activity on the schoolyard at the time. This meant that the effects of the size of the observation zones and other unwanted variables that could affect the observations should not have much impact on the patterns that show up on the map.

Devising symbology to show all the feature of the schoolyard was challenging. The maps needed to include the boundaries of the observation zones as well as a lot of information on the schoolyard, including constructed features, markings, sand boxes, and other surface types and vegetation, such as trees. The main focus of the map was to highlight the patterns in physical activity, and so that part of the map had to be prominent. Our solution was to symbolize the features of the map using textures that let the fill color, representing the data, to show through. The values on the map represent the portion of children who were engaged in moderate to vigorous physical activity (MVPA) as a portion of all the children observed in each observation zone. We chose a bi-chromatic scheme to symbolize activity, with red areas representing low relative activity levels and blue areas showing high physical activity levels. We symbolized the areas that were "average"—consistently similar to the physical activity in space as a whole—as light gray. This color scheme helped to direct the readers to the parts of the schoolyard that had an impact, either to encourage or discourage physical activity in children. An example of one of the maps is provided in Figure 12.6.

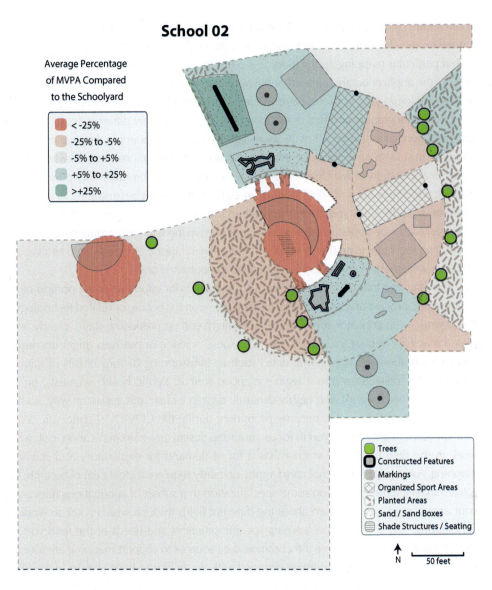

FIGURE 12.6
A map of schoolyard activity from a school in Denver, Colorado.

Engaging with the cartography community

As I have mentioned throughout this book, the field of cartography is currently experiencing enormous, unprecedented, and rapid change as new technologies and software applications continually push the bounds of mapping. Being a modern professional cartographer, therefore, truly requires one to be a *life-long learner*. After you become familiar with a particular mapping application, it is often merely a matter of time before new versions of the application are released or the application begins to slide into obsolescence as new technologies take its place. Beyond the fundamental ideas and concepts discussed in this book, cartographic practice draws from a nearly completely different set of technical skills and applications used 10 or 20 years ago. By way of an example, I put years of effort and time into learning to produce interactive maps and animations in Shockwave format using an application called Director (originally a Macromedia product, later purchased by Adobe), which was formally discontinued in 2017. While it can be frustrating to see a field move beyond the skills you have practiced and honed, the good news is that familiarity with many software applications translates well to pave the way for learning new ones. The underlying logic and even the "feel" of related software applications are often similar, and I believe we benefit from our experiences with the outdated ones.

You may find that your work directs you to focus your efforts on some portion of the cartographic process or a type of map, leading attention to specific or related technologies. If you work as an educator, for example, you might end up producing static, clearly articulated maps for web pages or texts. Work in social science or business might encourage you to learn about wrangling "big data" such as transforming billions of bits of data on social media interactions into a legible mapped format. Public health scientists, on the other hand, need to work with highly dynamic data in a clear and impactful way, such as one of the Internet maps that became prominent during the COVID-19 pandemic. Work as a journalist might require you to focus your map design as a communication tool, while work in the field of climate science has a lot of demand for exploratory and scientific mapping. As noted, professional cartography certainly requires a great deal of breadth, but it often makes sense to develop some specialization in a subset of applications relevant to your area of work. As you learn about the different fields through which you can work on cartography, you can focus on learning specific practices and needs of the field you are working with, as well as where the common data sources to support mapping are found.

An important facet of becoming engaged with the field is to keep abreast of the kinds of maps being produced. As is the case with writing or art, cartography is subject to an evolving set of ideas, principles, and trends (even some "fads"), as culture and society reacts to the eroding limitations of technologies. An easy way to check out what is going on is by checking out cartography blogs (try simply searching the Internet for "cartography blog," which typically yields remarkably prolific ranked blog lists). There are countless websites and blogs dedicated to highlighting excellent cartography, sharing original cartographic work, or discussing the latest trends and developments within the field.

Getting to know individuals working in the field ("networking") is often essential for gaining a foothold in the field. A good starting point is to join or become involved with professional cartography societies, whose missions are to promote cartography and support the professional efforts of its members. Several well-established societies hold annual cartography competitions for both professionals and students, where you can try to get your work recognized by competing for an award. Appendix 4 lists a sample of some prominent professional cartography societies.

The most important step you can take to start building your cartographic skill set and acumens, of course, is to start producing your own maps. If you are taking a cartography course, consider using your work to start building a cartography portfolio, a collection of your best map work. You can use the work to demonstrate the merits of your own cartographic design and technical ability to potential employers or individual clients. A huge amount of mapping now comes from amateur cartographers not associated with any organization, and, depending on your professional network, it is often not difficult to find people who need a good cartographer. Given the availability of affordable or open-source (free) software, it is entirely possible to produce an impressive array of maps if you have basic computer hardware and software, time, patience, and some measure of passion and enthusiasm for the work.

Conclusion

Many observers have dubbed the period we live as the *Information Age*. The rise of computer technology and ever-growing access to that technology have meant that a lot of power, economic activity, and policymaking is based around information. The importance of data and information in our society has infused mapping with more power than ever before, engendering the field of cartography with a great deal of salience, both as a professional skill and as force that affects society and policy. It is my firm belief that the study of cartography will continue to serve as an important tool for a variety of professional careers and that maps will always be a meaningful tool.

My goal in this book was to provide readers with some of the foundational knowledge and skills needed to produce good cartography. I hope the book conveyed the essential concepts in mapping, exposed you to some of the key tools and techniques you need to build your own maps, as well as imbued an appreciation for the immense power of maps. Just as importantly, I hope that, as you continue to examine, think about, and work with cartography, you discover the potential beauty, power, and intrigue of maps.

Discussion questions

1. Think of some topic you are interested in exploring and imagine that you were tasked with devising a printed, five-to-seven-page atlas on the topic. Make sure that you clarify the audience and potential readership of the atlas. What maps

would you include in the atlas? How would you structure it? What specific meas-ure could you take to ensure that the maps communicate clearly and are otherwise broadly accessible?

2. Take any of the mapping examples covered in this chapter and critically review the maps. What improvements or changes to the maps would you recommend? Can you identify any of the design principles discussed in the book that could be applied toward improving them?

3. Take some time to find some job listings seeking people with skills in mapping, GIS, or cartography. What areas of work are the job listings in? What kinds of skills or experience would you need to be competitive for the position?

4. A fundamental question early in this book, and one that cartography scholars have discussed for years, is whether the field is an "art" or a "science." Some scholars argue that placing that dualism on the field is limiting (see Krygier 1995) or false altogether. What is your opinion: is cartography more of an art or a science? Per-haps you think it is neither, or perhaps both?

References

Anthamatten, P. 2015. "Geography and Health." In *Health Studies: An Introduction (Third Edi-tion)*, edited by J. Naidoo, and J. Wills, 196–227. London: Palgrave.

Anthamatten, P., and H. Hazen. 2015. "Changes in the Global Distribution of Protected Areas, 2003–2012." *Professional Geographer* 67 (2):195–203. doi: 10.1080/00330124.2014.921014.

Johnson, S. 2006. *The Ghost Map: The Story of London's Most Terrifying Epidemic–And How It Changed Science, Cities, and the Modern World*. New York: Riverhead Books.

Krygier, J.B. 1995. "Cartography as an Art and a Science." *Cartographic Journal* 32 (1):3–10. doi: 10.1179/caj.1995.32.1.3.

Olson, D.M., E. Dinerstein, E.D. Wikramanayake, N.D. Burgess, G.V.N. Powell, E.C. Under-wood, J.A. D'Amico, I. Itoua, H.E. Strand, J.C. Morrison, C.J. Loucks, T.F. Allnutt, T.H. Rick-etts, Y. Kura, J.F. Lamoreux, W.W. Wettengel, P. Hedao, and K.R. Kassem. 2001. "Terrestrial Ecoregions of the Worlds: A New Map of Life on Earth." *Bioscience* 51 (11):933–938. doi: 10.1641/0006–3568(2001)051[0933:TEOTWA]2.0.CO;2.

Simon, G. 2016. *Flame and Fortune in the American West: Urban Development, Environmental Change, and the Great Oakland Hills Fire*. Berkeley: University of California Press.

World Database on Protected Areas. 2012. "Discover and Learn about Protected Areas." Accessed May 27, 2015. http://www.protectedplanet.net.

Map gallery: "Maps from the wild"

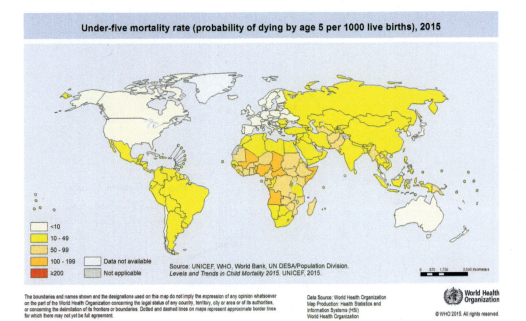

Under-five mortality rate (probability of dying by age 5 per 1000 live births), 2015

| <10 |
| 10 - 49 |
| 50 - 99 |
| 100 - 199 | Data not available |
| ≥200 | Not applicable |

Source: UNICEF, WHO, World Bank, UN DESA/Population Division.
Levels and Trends in Child Mortality 2015. UNICEF, 2015.

The boundaries and names shown and the designations used on this map do not imply the expression of any opinion whatsoever on the part of the World Health Organization concerning the legal status of any country, territory, city or area or of its authorities, or concerning the delimitation of its frontiers or boundaries. Dotted and dashed lines on maps represent approximate border lines for which there may not yet be full agreement.

Data Source: World Health Organization
Map Production: Health Statistics and
Information Systems (HSI)
World Health Organization

World Health Organization

© WHO 2015. All rights reserved.

FIGURE A.1
Map of mortality rates by country. World Health Organization. (September 9, 2015).
"World: Under-5 mortality rate (probability of dying by age 5) per 1,000 live births, 2015."
Retrieved January 21, 2018, from http://gamapserver.who.int/mapLibrary/Files/Maps/Global_
UnderFiveMortality_2015.png.

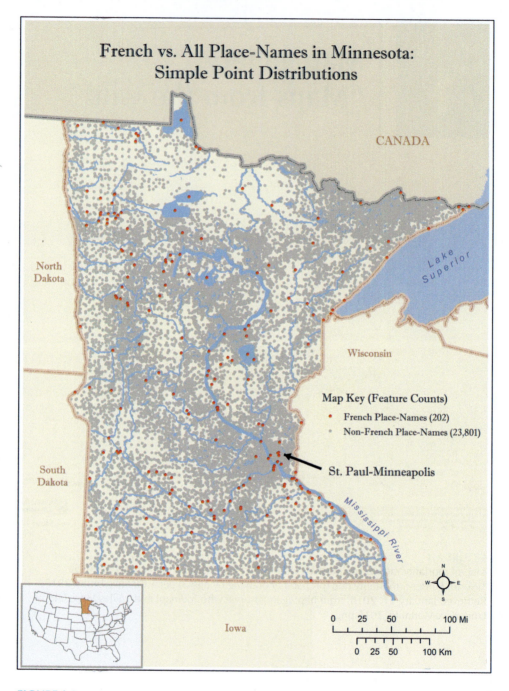

FIGURE A.2

Map of French place names and French ancestry in Minnesota, USA (Marcelle Caturia).

Source: Marcelle Caturia, Master's Thesis, University of Colorado Denver.

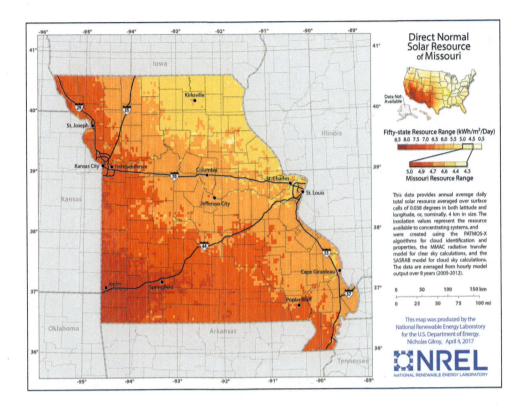

FIGURE A.3

Direct normal solar resource of Missouri (National Renewable Energy Laboratory).

Source: National Rewable Energy Laboratory, US Department of Energy (https://www.nrel.gov/gis/solar.html).

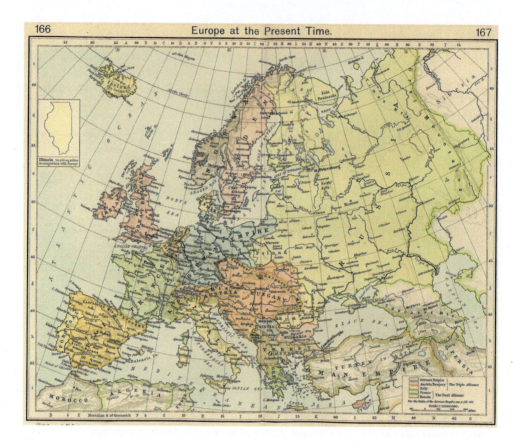

FIGURE A.4

"Europe at the present time".

Source: Shepherd, W. R. (1923). *Historical Atlas*. New York, Henry Holt and Company.

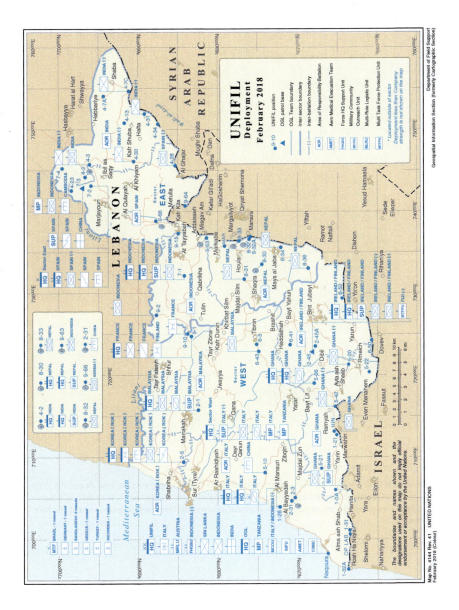

FIGURE A.5

United Nations Interim Force (UNIFIL) in Lebanon deployment, February 2018.

Source: United Nations. UNIFIL deployment February 2018 [cartographic material] " https://digitallibrary.un.org/record/1473371/files/4144Rev-41.pdf.

FIGURE A.6

Excerpt from a UK Ordnance map.

Source: An example of a UK Ordnance Survey Map, designed for 1:250,000. This map shows the area around the Wash in the North Sea. This contains OS data © Crown copyright and database right (2010). Printed under the Ordinance Survey OpenData liscence: https://www.ordnancesurvey.co.uk/business-and-government/licensing/using-creating-data-with-os-products/os-opendata.html.

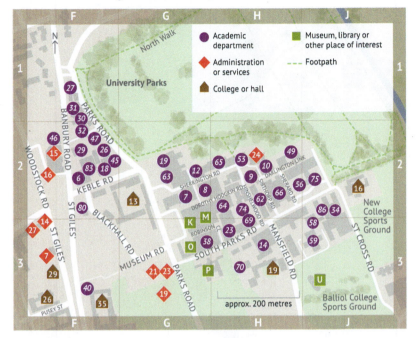

FIGURE A.7

A map of university buildings, Oxford University.

Credit: Sandina Miller / Oxford University Public Affairs.

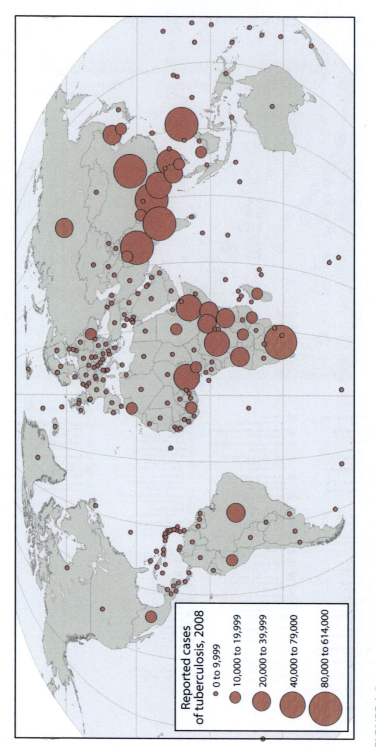

FIGURE A.8
Reported cases of tuberculosis.

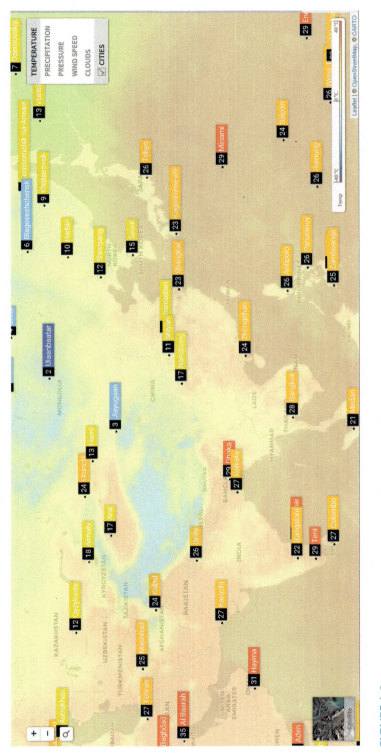

FIGURE A.9

Weather map of Asia.

Source: https://openweathermap.org/weathermap?basemap=map&cities=true&layer=temperature&lat=30&lon=-20&zoom=5.

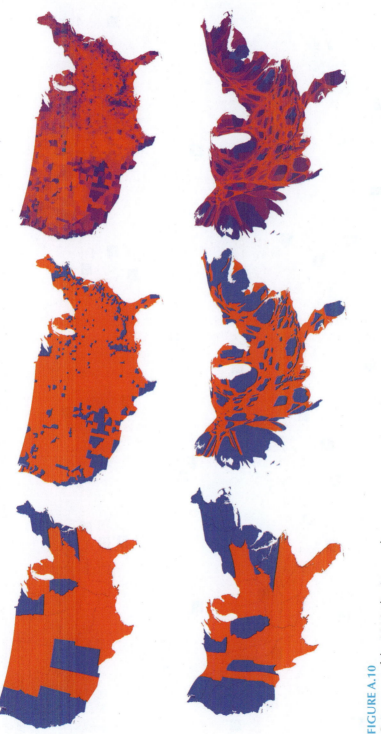

FIGURE A.10

Cartograms of the 2016 election results.

Credit: Mark Newman, Department of Physics and Center for the Study of Complex Systems, University of Michigan. (http://www-personal.umich.edu/~mejn/election/2016/).

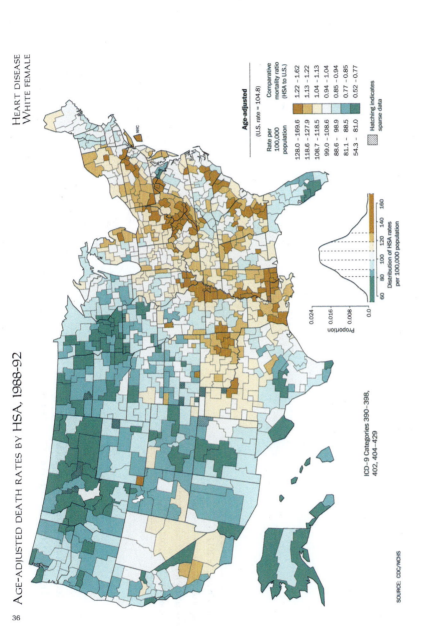

Heart disease
White female

Age-adjusted death rates by HSA, 1988–92

Age-adjusted

Rate per 100,000 population	Comparative mortality ratio (HSA to U.S.)
(U.S. rate = 104.8)	
128.0 – 169.6	1.22 – 1.62
118.6 – 127.9	1.13 – 1.22
108.7 – 118.5	1.04 – 1.13
99.0 – 108.6	0.94 – 1.04
88.6 – 98.9	0.85 – 0.94
81.1 – 88.5	0.77 – 0.85
54.3 – 81.0	0.52 – 0.77

Hatching indicates sparse data

ICD–9 Categories 390–398, 402, 404–429

SOURCE: CDC/NCHS

FIGURE A.11

A map from the *US Atlas of Mortality.*

Source: Pickle, L. W. and National Center for Health Statistics (1996). *Atlas of United States Mortality.* Hyattsville, MS, National Center for Health Statistics, Centers for Disease Control and Prevention, U.S. Dept. of Health and Human Services.

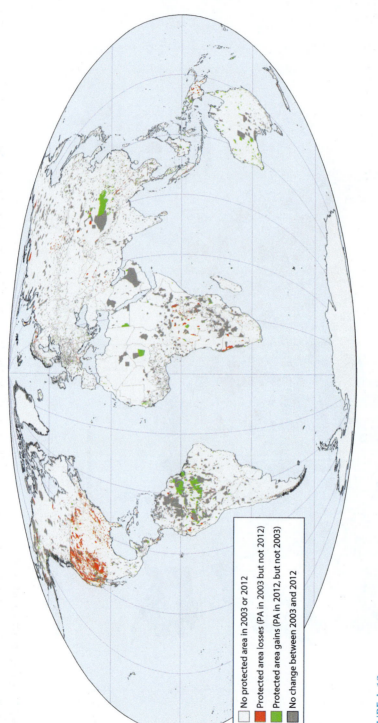

FIGURE A.12

World map of protected area gains and losses between 2003 and 2012.

Source: Anthamatten, P. and H. Hazen (2015). "Changes in the Global Distribution of Protected Areas, 2003–2012." *Professional Geographer* 67(2): 195–203.

Legend:
- No protected area in 2003 or 2012
- Protected area losses (PA in 2003 but not 2012)
- Protected area gains (PA in 2012, but not 2003)
- No change between 2003 and 2012

APPENDIX 2

Sources of spatial data

This appendix contains an annotated list of data sources that are useful for cartography. As the Internet seems to have been in an ever-changing state of flux since its inception, some of these may change, evolve, or disappear over time, and there are certainly large quantities of other data available. However, this list contains a list of several sources of data that have been excellent resources for work in geospatial science and cartography for years or decades. Many of these have well-established permanence on the Internet and are likely to be relevant for a long time.

Copernicus Open Access Hub (European Space Agency): https://scihub.copernicus.eu/

The Copernicus Open Access Hub was previously known as the Sentinels Scientific Data Hub and offers access to public domain, high-resolution imagery from the Sentinel satellites. It offers ten-meter resolution of satellite data with a user interface similar to the US counterpart (Earth Explorer). For users who are interested in remote sensing applications, it also has data on spectral bands.

Earth Explorer: https://earthexplorer.usgs.gov/

Earth Explorer is funded and maintained by the United States government and contains a rich variety of public domain, geospatial data. A strength of the website is the availability of raster data, such as high-resolution orthoimagery, digital elevation data, land cover data, among a rich variety of other data. While much of the data are focused on the USA, the website also offers worldwide data on some topics, such as Shuttle Radar Topography Mission (SRTM) digital elevation data. The website can be a little difficult to learn to use, but it offers a spectacular resource for cartographers, particularly if you wish to use raster data.

ESRI Open Data: https://hub.arcgis.com/pages/open-data

ESRI, the makers of *ArcGIS* software, stands to benefit from facilitating access to open-source data. In 2017, it launched an open-data portal in collaboration with hundreds of organizations. The data are linked to their original sources, which are continually updated. While this portal is not as focused as many of the others and will require some careful work sorting through data, it is worth visiting if you are seeking a specific type or topic of open-source data.

GeoNetwork (United Nations Food and Agricultural Organisation): http://www.fao.org/geonetwork/srv/en/main.home

This website from the United Nations is centered on the missions of the Food and Agricultural Organization but offers a much larger variety of data to achieve its missions to "improve access to and integrated use of spatial data information" and to "promote multidisciplinary approaches to sustainable development." It contains excellent, international basemap data in addition to topics such as agriculture and food, human health, water resources, climate, soil, and topography. Cartographers interested in sustainability on a global scale will find a great deal of useful data on this website.

Integrated Public Use Microdata Series (IPUMS) Terra: https://www.terrapop.org/

Social data are often difficult to find for international work, since so much of social data are driven by individual states, which often result in data that are not comparable and generally disjointed. The IPUMS Terra projected addresses this problem by providing a website that seeks to integrate census data from multiple countries. One of its missions is to disseminate microdata, which describes characteristics of families and their households. It also contains spatial data files on census and administrative units for much of the planet, as well as environmental data.

Natural Earth: http://www.naturalearthdata.com/downloads/

The website describes itself as "a public domain map dataset available at 1:10, 1:50, and 1:110 million scales. Featuring tightly integrated vector and raster data, with Natural Earth you can make a variety of visually pleasing, well-crafted maps with cartography or GIS software." A true benefit of the website is that it was constructed with cartographers in mind and contains well-organized data with suitable quality for small-scale mapping.

Open Street Map: https://www.openstreetmap.org/

As noted in the text, Open Street Map played a critical role in the democratization of geospatial data. It was started by British entrepreneur in 2004, in response to the restrictions placed on much of the spatial data available at the time. The data are compiled from a variety of sources, notably from social media data, such as Twitter, and from volunteered geographic data. You may have to spend some time working through third-party applications to use the open street data with your particular flavor of GIS software, but it is possible to use the data both with proprietary (such as ArcGIS) and open-source (such as QGIS) geospatial software.

Open Topography: https://opentopography.org/

The mission of Open Topography is to provide open access to high-resolution geographic data, which includes LIDAR and other remote sensing technologies. While it mostly covers the United States, the website has begun expanding to other countries and has "community dataspaces" with data on locations around the world.

Perry-Castañeda Library Map Collection: https://legacy.lib.utexas.edu/maps/

As the Internet was much younger, one of the first developments in making maps and mapping more accessible was the availability of static map images, and the Perry-Castaneda Map Collection provided excellent examples of historical and open-source maps. The maps are available in.pdf and.jpg format and include digitized maps from a variety of sources. If you wish to incorporate these into a GIS application, you will have to manually digitize them.

Sanborn Insurance Maps, US Library of Congress: https://www.loc.gov/collections/sanborn-maps/about-this-collection/

"The Sanborn Fire Insurance Maps Online Checklist provides a searchable database of the fire insurance maps published by the Sanborn Map Company housed in the collections of the Geography and Map Division." As noted in the text, Sanborn Insurance maps offer detailed information on cities in the USA during the nineteenth and twentieth centuries. The maps are images and require digitization to be used in a GIS.

The United States National Map: https://nationalmap.gov/

The National Map, published by the United States Geological Survey, provides a treasure trove of public domain, geospatial data. The website was formed as a result of a collaboration between the USGS and other government agencies to improve the availability of US geospatial data. From this website, one can view and download topographic maps (both current and historical), hydrological data, political boundaries, orthoimagery, transportation, elevation, geographic names database, among others. Particularly if you are interested in working on data in the United States, this is an excellent resource that offers high-quality spatial data.

United Nations Environmental Data Explorer: http://geodata.grid.unep.ch

This website is the authoritative data source used by the United Nations Environment Programme for its environmental assessment work. It is a spectacular public-domain resource for international work on the environment, containing information on topics such as endangered species, agriculture, forests, biomass, and political work around conservation. The website also offers a variety of non-geospatial data relevant to the environment, including databases, figures, and graphs. A strength of the website is its breadth of data, which cover a huge variety of topics and regions, as well as the quality of the data and the credibility of its source.

Eleven guidelines for constructing and critiquing maps

As you build a map project and review these guidelines, always bear in mind the fundamental (1) purpose, (2) audience, and (3) medium of the map.

(1) **proper use of map elements**: Do the map elements add to the message and aesthetic appeal to the map? Are there map elements that you think contribute nothing to the map's goals and could be removed? Is there information that you think is needed for the map to be used in its full capacity? Can map elements be added (such as a legend) to explain something that is not inherently clear?

(2) **choice of projection**: Are there any obvious problems with projection? Does the map or the graticule (if there is one) look odd in such a manner that it might interfere with the aesthetic appeal or the use of the map? Should shape or area be preserved in the map?

(3) **legibility**: Are all parts of the map, including text and symbology, easy to read?

(4) **the use of visual hierarchy**: Are the appropriate elements emphasized and de-emphasized? Is there something the map author can or should do to highlight or reduce the visual prominence of map elements? Can you reduce the visual weight of ancillary elements, such as county borders?

(5) **balance**: Does the layout look lopsided or do you notice gaps in the design? What could you do with map elements to convey better balance?

(6) **figure-ground and contrast**: Does the key feature of the map "stick out"? Can you change things to make the map more prominent? For example, could the author gray out the background, make the map completely enclosed by negative space, or give the mapped area a vignette?

(7) **correct symbolization**: If the map should convey order in the data, does it use the appropriate visual variables? If it uses a non-traditional visual variable for rankable data (such as hue), is the order clear? Are points, lines, and areas used appropriately, or could the features on the map be symbolized differently or better?

(8) **intelligent use of text and typography**: Is the prose well-written and spelled correctly? Do labels interfere with map features or other labels on the map? Does the map utilize typographical options (such as font, tracking, font style) well?

(9) **intelligent use color**: Do the colors work well together? Do readers find the colors annoying? Have colors been selected that should not make it difficult for color-impaired people to read? Can you read and interpret everything well?

(10) **Tufte's principles of graphic design (such as unnecessary elements on the map or a favorable data-ink ratio)**: Is there *mapjunk* (stuff on the map that adds nothing to the message or the visual appeal? Can data (e.g., map features) be removed or generalized to more effectively convey the message of the map? How can the data-ink ratio be improved? For example, if the map shows political spending by county, are rivers necessary? Could only a few rivers be shown, or could you remove some "non-data ink"?

(11) **the overall appeal of the map**: Is the map visually appealing? Is there interesting material on the map to give it a captivating design? What irritates you about the map and what do you like? At some point, you must recognize that there is an *art* to making an excellent map; think about how you can draw inspiration from art and other excellent maps. Art historians and critics can learn to recognize good art after spending hundreds of hours viewing and thinking about art. As a cartographer, you should have experience viewing and thinking about maps and should come to recognize good mapping when you see it.

What can be done to the map to nudge the map toward "cartographic excellence," beyond the guidelines provided here?

APPENDIX 4

Professional cartography societies

TABLE A4.1 This table shows some of the major national and international cartography societies, generally dedicated to supporting promoting collaboration among professional and amateur cartographers. The text in the tables was quoted directly or translated from the societies' websites

Name	Goals and Aims	URL
Australian and New Zealand Map Society Incorporated	"The general aims of ANZMapS are: To promote communication between producers, users and curators of maps, to improve the skills and status of persons working with map collections, [and] to promote the development and effective exploitation of map collections throughout Australia."	anzmaps.org
British Cartography Society	"A dynamic association of individuals and organisations dedicated to exploring and developing the world of maps."	cartography.org.uk
Canadian Cartographic Association	"The CCA is a non-profit association created to promote the disciplines and professions of cartography and Geographic Information Science in Canada."	cca-acc.org
Cartographic and Geographic Information Society (CaGIS)	"CaGIS provides an effective network that connects professionals who work in the broad field of Cartography and Geographic Information Science both nationally and internationally."	cartogis.org
Deutsche Gesellschaft für Kartographie e.V. (DGfK)	(*German-speaking*) The DGfK supports cartographic and geospatial information in research and practice. It is dedicated to furthering professional development and education and preserving cartographic culture in collaboration with national and international cartographic societies.	dgkf.net

(Continued)

TABLE A4.1 *Continued*

Name	Goals and Aims	URL
International Cartographic Association/Association Cartographique Internationale (ICA/ACI)	"…the mission of the International Cartographic Association (ICA) is to promote the disciplines and professions of cartography and GIScience in an international context."	icaci.org
International Map Industry Association	"The International Map Industry Association (IMIA) is dedicated to fostering the growth of the mapping industry through the exchange of ideas and information, the education of industry trends and technologies, and access to a wide spectrum of mapping leaders."	imiamaps.org
North American Cartographic Information Society	"NACIS was founded in 1980 and is comprised of specialists from private, academic, and government organizations along with people whose common interest lies in facilitating communication in the map information community."	nacis.org

Glossary

accuracy: the extent to which data or a measurement is true

additive color theory: color mixing that is based on adding luminance of different colors

additive primary colors: the primary light colors used for color production in digital media: red, blue, and green

age adjustment: the statistical manipulation of population data to remove the effect of age structure from variance in mortality data

aggregation: a generalization operation that groups multiple point locations into a single area symbol

allocentric: from a third perspective, different from one's own; allocentric spatial thinking is the ability to view one's self in the context of a spatial representation, such as a map

amalgamation: a generalization technique that combines several area symbols into a single area symbol

anaglyphs: a stereoscopic image that consists of the superimposition of two images in different colors that mimic the left-and right-eye perspectives, designed to produce the illusion of three dimensions when viewed through colored optical filters

analysis: one of the foundational abilities of a GIS; the manipulation or processing of data to produce new knowledge or information that would not have existed with the data alone

animated cartography: a form of cartography that involves motion or dynamic alteration of map symbols; the application of animation to add a temporal component to maps

anti-aliasing: the practice of blending colors on computer screens so that the text appears seamless with a continuous background, rather than appearing blocky

Application Programming Interface (API): a system of protocols and processes in an operating system that enables users to develop software applications

area symbols: cartographic symbols that appear as areas or polygons on a map; area symbols are often used to show political territory, forest coverage, classes of geology, or other features with an aerial extent

aspect: (projections) the angle at which the developable surface aligns with the earth; projections are classified as normal, oriented along the earth's north-south axis, transverse, oriented around an east-west axis; or oblique, neither normal nor transverse

aspect ratios: the relation between the width and height of a digital display

atmospheric attenuation: the simulation of atmospheric haze, such as one might view over a large area, to produce the visual effect of depth

attribute data: a GIS term used to refer to data attached to a spatial feature

authorship: an element on a map layout that indicates the cartographer who produced the map

automated cartography: computer models that assist with making cartographic decisions, such as how much to select or generalize features and place labels

automated generalization: computer operations that can perform generalization following strict operational rules

azimuthal (map projection): projections in which distance is preserved from the center of the map to all other points

balance: the distribution of map elements across a layout

base map: the features of the map that serve as the background or context

beveling and embossing: A text effect that gives it the appearance of being slightly extruded

blurring: a graphical operation that makes the boundaries between distinct colors blurrier and less clearly defined

bmp (file format): A graphical file format developed by *Microsoft* for use in *Windows* applications

bold/boldface: a font typeface with thicker line

brightness: the amount of radiation emitted from luminous objects, such as a computer screens

camera aspect: the angle at which a visual perspective is oriented

Cartesian coordinate system: a two-dimensional coordinate system that identifies the location of a point with a coordinate pair that specifies the distance of the point from a horizontal x-axis, and a vertical y-axis

cartogram: a type of map that distorts the areas and shapes of the areas to show the relative importance of each of the units in terms of some attribute

cartographic generalization: the graphical manipulation of symbols on a map to make them legible at the map's target scale

cartographic scale: the relation between distance on a map and the actual distance it represents on the earth. A small-scale map represents a large amount of the earth's surface on a small amount of map space, such as a page-sized map of the world (e.g., Figure A.1). A larger-scale map represents a relatively small amount of the earth on the same map space, such as a page-sized or larger map of a university campus (e.g., Figure A.7)

cartography: the study and practice of the art and science of mapmaking

case (projection): the case of a projection refers to whether it is a tangent projection, touching the modeled surface of the earth at a single point or standard line, or a secant projection, in which the model touches the earth in a multiple points or lines

category range: the highest and lowest values in a category of data

chartjunk: a concept coined by Eduard Tufte to refer to "overly-busy grid lines and excess ticks, redundant representations of the simplest data, the debris of computer plotting, and many of the devices generating design variation" (Tufte 2001, 107)

choropleth maps: maps that use color or shading to show data about statistical-administrative units (such as countries, states, counties, and provinces)

cloud computing: the use of a remote server, rather than a local computer, to perform computing functions, such as the storage, management, processing, and analysis of data

CMYK: the main colors used in printing and the basis of subtractive color schemes: cyan, magenta, yellow, and key (black)

coalescence: a cartographic problem requiring generalization that occurs when features are so close that they appear to touch, when they are in fact distinct objects

cognitive lag: the time it takes to mentally process a novel map interface

cognitive load: the amount of information users must hold in memory to comprehend a map or make meaningful comparisons

collapsing: a generalization technique that replaces an object's physical details or shape with a symbol

color cycling: the rapid succession of colors along a path to simulate motion

color hue: the appearance of a color as determined by its dominant wavelength; "blue," "purple," and "turquoise" are examples of color hues; color hue is normally appropriate as a visual variable for representing qualitative data

color saturation: the intensity or purity of a color; colors that lack saturation appear as grayscale; color saturation is a quantitative visual variable

color value: the lightness or brightness of a color; color value is a quantitative visual variable

color vision deficiency: the inability to distinguish between certain shades of color, more commonly referred to as "color blindness"

communication: the practice of transmitting information from one individual to another; one of the key purposes of maps is to communicate spatial information

compartmentalized maps: online maps with graphical separation of the map and the other cartographic elements

compass rose: a map element that indicates the orientation of the cardinal directions on a map

completeness: a facet of data quality that refers to how well the objects in a database represent the universe it is purports to represent

complication: the idea that the generalization process depends upon conditions in different areas of the map

compromise (map projection): a map projection that preserves neither shape nor area

cone cells: pigmented optical cells in the eyes that absorb visible light radiation and send signals to the brain in response

conflict: a cartographic problem requiring generalization that occurs when the symbols on a map communicate an idea that is logically inconsistent with the rest of the map

conformal: a map projection that preserves shape but severely distorts area

congestion: a cartographic problem requiring generalization that occurs when too many features appear in too small a space

conic projection: a family of projections that is developed from a developable surface shaped like a cone

continuous data: data for which there is some value at every point in space; temperature, elevation, and concentrations of air pollution are good examples of continuous data

contour lines: lines on a map that join points of equal values in a continuous distribution; topographic maps often contain contour lines that show elevation

coordinate system: a systematized method for identifying location with numbers

counter-mapping: a localized, participatory, and anti-technocratic form of cartographic practice which draws on existing experiential knowledges to generate alternative, counter-hegemonic representations of geographic space

critical break: a point in a data set at which there is some important distinction between the data ranges

critical cartography: cartographic study which scrutinizes the linkages between mapping, geographic knowledge, and various forms of power

crowd-sourced data: data produced by volunteers

crude rate: a simple rate that is not adjusted for additional variables; the term is often used in maps of health to distinguish between a simple mortality rate, in contrast to an age-adjusted rate, which accounts for the age structure of a population, in addition to the population counts

csv (file format): csv stands for "comma separated values," a simple file format for storing tabular data; due to the simplicity of the data structure, csv files can be used in a broad range of applications designed to read tabular data

cybercartography: see *web mapping*

cylindrical projection: a family of projections that is developed from a developable surface shaped like a cylinder

database join: the process of linking different database tables using a key field, a common attribute; database joins are useful for linking features in a GIS to attribute data from a table

data capture: the transformation of real-world information into a form that a GIS system can manipulate

data certainty: the degree of confidence in the accuracy of data

data classification: the methods for classifying data into groups

data consistency: the absence of conflicts in a database, referring to the logical integrity of the database architecture

data density: a term coined by Eduard Tufte to refer to the "number of entries in a data matrix" divided by the "area of the data graphic" (Tufte 2001, 162)

data-driven: scientific research and methods designed to accommodate and work around available data; a data-driven question is framed by available data

data error propagation: the propagation of error through transformations of data; geospatial data must often undergo multiple rounds of data processing including collection, digitization, and transformation which makes the data vulnerable to error propagation

data-ink ratio: a term coined by Eduard Tufte to refer to the amount of data ink—ink dedicated to communicating variations in the data—compared to non-data ink, the rest of the ink on the map that does not communicate data

data lineage: the history of the data from its original collection and its transformation along the way to its current form; geospatial data often include information about the data lineage

data sources: a map element that indicates the sources of the data on the map

data storage: one of the key facets and abilities of modern GIS; the ability to store lots of spatial data in the form of digital files

dbf (file format): "database files"; a common format for storing tabular data

decimal: a base 10 numerical system, the prevailing number system

decorative fonts: special fonts designed to convey a theme or emotion

defined interval: a data classification scheme that uses some *a priori* division of data, often used to build clean-looking or "pretty" breaks in the data

deuteranopia: the inability to distinguish between colors in the green, yellow, and red portions of the visible spectrum

developable surface: a three-dimensional shape that can be flattened into a plane without any distortion or tearing

digital raster graphic (DRG): a digital image from scanning a map

digitization: the process of converting the symbols on a paper map or scanned digital image into georeferenced features in a GIS

digitizing tables: tables designed for digitizing maps

dimensionality: used in cartography to refer to the number of dimensions in cartographic symbols: points, lines, areas, and volumes

diode: tiny lights that comprise a computer pixel; pixels are normally comprised of a red, blue, and green diode

discourse analysis: a field of study in the social sciences which examines different forms of naturally occurring discourse (such as the popular media and speeches from politicians) to gain insight about the structures and underlying assumptions of society

discrete data: data that have a clearly defined starting and end point, a presence or absence

displacement: a generalization technique that involves moving a map feature so that it does not interfere with other nearby map features

(map) display: the term is often used to refer to a temporary map image, in contrast to a map designed to be a refined means of communicating spatial data

diverging color scheme: a color scheme that uses different hues to show sequential data; the most effective examples use one hue to represent values below a mean or standard, and a highly contrasting hue to represent values on the other end of the spectrum with values above a mean or standard

dots per inch (dpi): the primary measure of the graphical precision of printers

drop shadow: a graphical technique that gives the appearance of a shadow around objects

drum printers/chain printers: computerized typewriters that make ink impressions with print characters

duration: a component of animated cartography that refers to the length of time a single frame is shown on an animation

dynamic cartographic generalization: generalization operations that are performed automatically or on-the-fly in response to user inputs such as zooming

dynamic maps: digital maps that change over time (such an animation) or in response to user inputs

dynamic text: text on a cartographic map that can change as a function of the scale of the map or other input variables; examples include text that changes size when a user zooms in or out on a map or text that only appears within a specified map scale range

egocentric: from the perspective of one's self

enhancement: a generalization technique that places emphasis or enhances a specific feature or component of an object, often with a communication objective in mind

eps (file format): "encapsulated post-script," a vector graphics file format with relatively simple encoding that can be used across a broad range of vector graphics applications

equal intervals: a method of data classification that classifies measurements into data categories with equal category ranges

equator: an imaginary great circle line around the earth that is equally distant from both poles

equidistant (map projection): a map projection that preserves distance from one or two points on the map to all other points along standard lines

equivalent (equal-area): a map projection that preserves area but severely distorts shape

exaggeration: a generalization operation to enhance the size of an object on a map to make it legible to the reader to or to draw attention to it

extensible: a file format that can be extended with additional information or functionality

extent: the spatial limits of a geographic area for study, mapping, or analysis

false easting: the "y-axis" in a projected coordinate system, from which distances to the east are designated; false eastings are most commonly applied in the Universal Transverse Mercator (UTM) coordinate system

false northing: the "x-axis" in a projected coordinate system, from which distances to the north are designated; false northings are most commonly applied in the Universal Transverse Mercator (UTM) coordinate system

false precision: the unscrupulous practice of using high precision to imply high accuracy

feature geometry: vector-based point, line, and polygon data in a GIS

figure-ground: the visual separation of figure, or foreground, from the background of a layout

fill (graphics): the color or pattern that occupies the interior portion of a vector-based graphical object

flipbooks: a metaphor for animated cartography that consists of a systematic progression of a series of maps over which the user does not have control, like an animated cartoon

flow map: a type of thematic map that communicates the flow rate of movement through variation in the width of the line symbols

fluid maps: online maps that do not use neat lines to separate the map from map elements or text and often occupy the entire web page

font: synonymous with *typeface*; the style and shape of the letters

foveation: foveated imaging is a technique in which the amount of detail provided varies as a function of where users fixate their gaze; foveation studies often use eye-movement trackers to determine users' viewing patterns on maps

frame line: lines around the entire layout

geodatabase (gdb): a file format that can contain geographic feature classes, data sets, rasters, and tables; the primary geospatial data format for vector data used in *ArcGIS*

geodesy: the study of the earth's shape; important for work to produce highly accurate mapping

geographic datum: a reference system modeled to approximate the shape of the earth's surface; cartographic projections are based upon a specific geographic datum

geographic positioning systems (GPS): a radio navigation system that relies on satellites to enable devices to indicate location

geographical coordinate system: a spherical coordinate system, originally conceived by Greek scholar Eratosthenes, that functions by specifying angles from the great circle plane formed by the equator (latitude) and the prime meridian (longitude)

geoid: "earth-like"; a term used to refer to the actual, slightly lumpy shape of the earth

geometric center: the physical center of a map

georeferencing: the process of attaching coordinates to data; "georeferenced data" refer to data that have spatial references

geospatial PDF: a class of pdf file that enables users to mark location and make map measurements with a user-specified coordinate and measurement systems

GeoTiff (file format): an extension of the Tiff format that allows georeferencing information; a properly formatted GeoTiff image can be loaded directly into a GIS and mapped

gif (file format): "Graphical Interchange Format"; a graphical file format constrained to eight-bit color and consumes relatively little file space

gnomonic: A projection developed from a model with a light source at the center of the earth

graduated symbol map: a type of thematic map that shows the magnitude of some theme or topic by varying the size of a point symbol

graphical scale bar: a type of map scale that graphically shows units of distance on the map in terms of the real-world distance

graticule: printed lines of longitude and latitude that comprise the key parallels and meridians in a geographic coordinate system on a map

great circle: the line formed by any cut that divides a sphere into equal halves

great circle route: the shortest route along the surface of a sphere between two points, which follows a great circle line that connects the points

grid north: the direction of north as indicated by the gridlines on a map

ground distance: the distance between two features on the earth's surface, often used in contrast to "map distance," which the distance between representations of those features on the surface of a map

heads-up digitization: digitization performed with a scanned image on a computer screen (in contrast to using a physical map and a digitizing table or tablet)

heatmap: a graphical representation of data that uses color to show the intensity or density of a geographic feature

hexadecimal: a base 16 numerical system used to build color codes in a variety of computer programming applications; in hexadecimal color codes, the first two symbols make up a number between 0 and 255 that refers to the red channel, the middle two symbols specify green channel, and the final two to the blue channel

hillshading: a technique used with relief maps that adds a subtle degree of shading to emphasize the shape of the topography

HSV color model: a color model built from the fundamental components of human color perception: hue, saturation, and value

impact printers: computer printers that function by impressing ink upon paper, like typewriters

imperceptibility: the idea that some geographic features are too small to be shown on a map when scaled to their actual size; when this occurs, exaggeration is needed to make the map legible

inconsistency: occurs when generalization decisions are not consistently applied across the map

ink-jet printers: printers that function by spraying an ink pattern on a sheet from small droplets controlled through electrical charges

inset: a small map on the layout that can serve either to clarify the context of the main map or to provide a larger-scale version of a detailed area of the map

interactivity: a feature of web and mobile cartography that enables users to interact with the map content

interrupted projection: a type of projection that tears, or interrupts, the map representation to reduce the severity of distortion

interval data: data which have specific numbers attached to them but lack any "true zero point," prohibiting some mathematical operations such as division or multiplication; temperature given as Fahrenheit is an example of interval data

isoline map: a map that uses contour lines, lines of equal value, to represent the distribution of continuous geographic phenomena

isopleth map: a map that uses contour lines, lines of equal value, to represent the distribution of continuous geographic phenomena; isopleth maps are occasionally distinguished from "isoline maps" by the practice of shading the areas between the isolines to represent a value

italics: a font typeface that tilts text to the right

Jenks Natural Breaks Algorithm: the default classification method in *ArcGIS* that uses a mathematical algorithm to determine natural breaks in a set of data

jpg (file format): "Joint Photographic Experts Group"; a raster graphics file format that uses a compression algorithm that preserves the quality of the image while reducing the file size

kerning: the adjustment of spacing between letters to accommodate specific letter combination

key: see *map legend*

key color: a component of the CMYK color model; the "K" refers to "key," or black ink

key field: a field in a table that can be used to link the data to other tables

key numbering: the use of a number (or some other succinct symbol on the map, such as a letter) with a listed and ordered reference outside of the map

kinetic effect: the perceptual phenomenon in which the three-dimensional structure of an object is more accurately perceived when the object or the viewer is in motion

laser printers: computer printers that function by reflecting upon a light-sensitive surface to build an electrostatic charge on the drum in a precise pattern, which is then inked on paper

latitude: the angle and direction on the surface of the earth from the equatorial plane

layout: the collection of graphical and written components to a map or display

leadering: a line that is added to connect the label with its target feature

legend ("key"): a map element included to explain the meaning of symbols that appear on the map

legibility: a fundamental principle of map design that calls for all of the textual and symbolic components of a map to be legible to the reader

light effects: graphical operations to produce photorealistic effects to images and other graphical objects

lightness: the amount of radiation the amount of reflected light from objects; lightness in color is synonymous with "value".

line spacing: the distance between lines of text

line symbols: cartographic symbols that appear as lines on a map; line symbols are often used to show rivers, roads, boundaries, paths, as well as the graticule (the lines of longitude and longitude)

longitude: the angle and direction of a point on the earth from the great circle formed by the prime meridian

lossless: a graphical file format that does not lose any graphical quality with compression

magnetic north: the direction toward which the needle of a compass points

major properties (map projection): the key properties that are distorted by a map projection: angle and area

map design: a key component of cartography around how to arrange a map, what map elements to include, and how to represent geographic data

map distance: distance on a map; often used to refer to the distance on a map between two specific points, often used in contrast to "ground distance," the distance between two features on the earth's surface

map elements: the individual components that make up a map layout; examples of map elements include the map, title, legend, and scale bar

map purpose: the overall function and intent of a map

map scale: a component of the map that communicates the cartographic scale; map scales can assume multiple forms, including a verbal scale, a representative fraction, or a graphical scale bar

mapjunk: excess ink on a map that does not contribute to the data story or the clarity of the map, such as excessive grid lines or reference ticks, using unnecessary graphical ornamentation on the data themselves, or overbusy political borders or symbology on a base map

masking: a graphical technique to block features to avoid graphical interference

merging: a generalization technique that combines several line symbols into a single line

meridians: lines of longitude

metadata: "data about data"; metadata is often included with data and includes information such as the data's origin, the time of its collection, and data lineage

metamorphosis: a technique in animated cartography that refers to the change in the shape of a feature over time

milliradians: A unit of measurement for angles, equivalent to about .0573 degrees

minor properties: the properties of distance and direction in a map projection

model and camera: a metaphor for animated cartography that places a user's perspective in a simulated three-dimensional environment

monospaced: fonts designed for computer applications with an equal spacing between each letter

natural breaks: a data classification scheme that divides data according to observed breaks in the distribution of the data

neat line: lines around the mapped area that separate the part of the layout that represents some portion of a mapped surface from other parts of the layout

negative space: the spaces and gaps between elements in a map layout; negative space is important to give a layout legibility and balance, synonymous with "white space"

nominal: a level of measurement that means "named," representing a quality of a phenomenon, rather than a measure or count; examples of nominal data include "language spoken" (e.g., English, German, Amharic, or Chinese) or "type of forest" (e.g., tropical, temperate, boreal)

non-impact printers: modern computer printers that print through non-impact methods; much modern cartography can be effectively printed with non-impact printers

north arrow: a map element that shows how the map is oriented by indicating the orientation of north

oblate spheroid: a slightly squashed sphere that is longer along one axis than the other; the earth is shaped like an oblate spheroid

ontology: the study of the nature of existence

open-sourced mapping: mapping software or data that anyone with access to the Internet and appropriate computing power is free to access and alter

opponent process theory: the idea that the eye contains three different types of cells that receive and respond to specific ranges of visible light frequencies; the key information the brain uses to perceive color is the *difference* in intensity between the signals

optical center: where people focus their attention on a map layout

ordinal data: data can be ranked into logical order, but it is not possible to perform mathematical operations on them. Examples of ordinal data include "village," "town," and "city," phenomena which have a clear and logical ordering, but upon which it is not possible to perform mathematical operations

orientation (visual variable): a visual variable achieved by varying the orientation of a point symbol, hatch marks in a line, or fill of an area, most appropriately used for qualitative data

orthographic: a projection developed from a model with a light source at an infinite distance

panning: a function in dynamic maps that enables the user to change the scale by zooming in or out

parallels: lines of latitude

pdf (file format): a stable, low-memory, flexible, and universally readable vector-graphics file format developed by *Adobe*

pixel: the smallest unit of programmable graphical space on a computer screen

pixels per square inch (PPI): a measure of the resolution of a digital display; the number of pixels within a square inch

planar coordinate systems: a general term for a coordinate system designed to specify locations in a two-dimensional plane; projected coordinate systems are planar

planar projection: a family of projections that is developed from a planar developable surface, such as an azimuthal projection

plotter printers/plotters: originally the term referred to vector-based printers that functioned with pens; the term now generally refers to large-format printers

png (file format): "portable graphics format"; a file format commonly used for images on the Internet, developed to work in a broad range of computer applications

point distribution map: a type of thematic map that displays how classes of points are distributed across an area

point symbols: a type of cartographic symbol that appears as a small point or symbol, often used to represent phenomena, such as cities and towns, or landmarks

pointer: a metaphor for animated cartography that refers to the practice of highlighting some portion of the map by adding lines, illuminating a part of the layout, or adding notation to the map

points (typography): a size measurement for text, equivalent to 1/72nd of an inch

popup window: a feature of some interactive cartography that presents information in a separate window

precision: the specificity of a measurement or value

primary colors: the "pure" colors from which all other colors are built: blue, red, and yellow.

primary data: data that were collected with a specific purpose or research project in mind

prime meridian: an imaginary great circle line around the earth that passes through the Royal Observatory in Greenwich, England

print resolution: the graphical resolution of a printer, normally quantified as dots per inch

projection (cartographic): the process of transforming the three-dimensional surface of the earth into a two-dimensional representation, such as a map on a sheet of paper; projection necessarily distorts some facet of that representation (particularly shape and area)

projection family or class: a group of projections with a common developable surface; examples include planar, cylindrical, and conic projection families

pseudo-3D (also called 2.5D): two-dimensional imagery that conveys a 3D perspective without the use of any form of stereography

qualitative data: a general term that refers to data referring to a quality of a feature rather than to any measure or count; qualitative lack any inherent logical ordering

quantiles: a method of data classification that places an equal number of measurements into each data category

raster data: a spatial data model that consists of a grid comprised of pixels, each of which contains an attribute value (e.g., temperature or elevation)

rate: a measure or quantity that is measured against some other measure or quantity (usually achieved through division)

rate of change: a term in animated cartography that refers to the rate at which a given feature changes on a map over a given period of time

ratio data: data on a count or measure of some phenomenon with a true zero, enabling a wide range of mathematical operations and manipulation of the data; population counts and disease rates are examples of ratio data

real-world distance: synonymous with "ground distance"

reference ellipsoid: an ellipsoid modeled to provide a coordinate system for an area being mapped

reference map: a general-purpose map not guided by any topic or theme; reference maps provide a general overview of a geographic region

refinement: a generalization technique that retains specific portions of a network to provide the reader with an idea of the characteristic and nature of the phenomenon being represented

relief models: a physical 3D model of the environment with raised terrain

remote sensing: the collection of geographic data through aeroplanes or satellites

representative fraction (RF): A method of reporting the scale of a map in the form of a ratio

RGB: the main colors used in digital displays and the basis of additive color schemes: red, blue, and green

rhumb line: line on a map that enables plotting a straight course with a constant bearing

rich Internet applications (RIAs): web-based applications that work like desktop software

rod cells: optical cells in our eyes that respond to the intensity of light signals

sans-serif: a family of fonts characterized by the lack of serifs, or finishing strokes

scale: refer to *cartographic scale*

science-driven: a research project that is driven and framed by the needs of science; the term is used in contrast to projects that are framed by the availability of data

screen resolution: the resolution of a digital display, given in pixels per square inch

script fonts: a family of fonts designed to mimic handwriting

secant projections: a type of projection in which the model touches the earth in multiple points or lines; the developable surface in a secant projection is modeled as passing through the surface of the earth

secondary data: data collected by sources other than the author or investigator of a project

selection: the removal or inclusion of a set of features to make a map legible, which must occur on any map; the term also refers to activating a part of a map, an element in a graphics program for manipulation

semiotics: the study of codes, signs, symbols, and their interpretation; semiotics can also include study of the social processes that make use of signs and symbols in order to communicate and produce meaning

serif: small finishing strokes on letters; serif fonts refer to a family of fonts characterized by the presence of finishing strokes

sexagesimal: a base 60 numerical system; angular degrees are divided into 60 minutes, which are further divided into 60 seconds, which is an example of a sexagesimal number system

shape (visual variable): a visual variable achieved by varying the shape of a point, line, or fill of an area, most appropriately used for qualitative data

sharpening: a graphical tool that makes the boundaries between distinct colors sharper and more clearly defined; sharpening functions as the opposite to "blurring"

shp (file format): one of the predominant file formats for storing vector data that can be read by a GIS

simplification: a generalization process of selectively reducing the number of points in a line to represent an object

simultaneous contrast: the way that perception of colors is affected by the presence of other colors around it

size (visual variable): a visual variable achieved by varying the size of a point, width of a line, or fill symbols in an area; size is a quantitative visual variable

slideshow: a metaphor for animated cartography that consists of a simple progression of a sequence maps in a way that gives a presenter or user control of the progression

slippy maps: dynamic digital maps that enable a user to pan around the map area

small circle: any straight line on the surface of a sphere that is not a great circle

smoothing: a cartographer generalization operation that helps smooth the "angularity" of line to form a smoother, more flowing representation

spatial accuracy: estimated agreement between the location of features on the map and their actual locations in the real world

spatial cognition: thought about the organization and manipulation of one's spatial environment

spatial data: information or data that are geographically referenced or which contain information about location

spatial data model: the operational form in which spatial data are stored and read by a GIS; dominant spatial data models are vector and raster data

spatial thinking: a collection of cognitive skills and mental processes that relate to thought about concepts of space and location

spectral signature: the unique combination of electromagnetic wavelengths that is reflected by a material

spherical coordinate systems: a general term for a coordinate system designed to specify locations on a sphere using angle measurements; the geographical coordinate system (based on latitude and longitude) is an example of a spherical coordinate system

stage and play: a metaphor for animated cartography in which components of a map are animated against a static backdrop, often to highlight a feature or show change over time

standard deviation: A statistical measure of the variability of data; the standard deviation data classification method groups data according to how far it falls from the group average, based on the standard deviation of the entire data set

standard line: the point at which a model of the earth's surface in a projection touches the surface of the developable surface; there is no distortion on the standard line

standardization: the practice of dividing a count of a phenomenon by some other value to remove the variance of the other value; population density is an example of standardized data because population counts are divided by the area; this technique is useful for mapping because it removes the unwanted effect of variance (such as area or population) of some feature of data

static map: a map that is not dynamic, with which a user may not interact; the term is normally reserved for digital maps, used to distinguish from maps that contain motion or interactive components

stereographic: a projection developed from a model with a light source from the opposite side of the earth from the area being mapped

stereopair: a pair of photographs designed to mimic the left- and right-eye perspectives of the subject; stereopair photographs can mimic the illusion of 3D viewing

stroke (graphics): the outline of a graphical object

subtractive color theory: the production of color through the process of selectively subtracting portions of the light spectrum by adding inks

subtractive primary colors: the primary ink colors used for color production in printed media: cyan, magenta, and yellow

svg (file format): "scalable vector graphics"; text-based, open-standard extensible vector graphics files

symbolization: the types of symbols used to represent geographic phenomena on a map

tactile maps: maps designed to be read through touch rather than through vision

tangent projections: a type of projection in which the model touches the earth in a single point or line of tangency

target feature: a feature to which a label is intended to refer

temporal accuracy: the "currentness" of the data, or how up-to-date the data are

text halo: text with an outline that is a different color from the text itself

text splining: the placement of text along a curved baseline

texture mapping: the graphical application of a pattern to a surface

texture (visual variable): the fill pattern of a point, line, or area; texture is an example of a quantitative visual variable

thematic accuracy: accuracy of the attribute data, the quality of information about the spatial features

thematic map: a type of map that is designed to communicate information about a topic or theme, such as infection rates or population density; this term contrasts with "reference map," a general-purpose map without any clear theme

Thiessen polygons: polygons whose boundaries define the area that is closer to a point than any other area

tif (file format): "Tagged Image File"; a lossless, extensible, graphics file format that often consumes a high amount of computer memory

time series: a series of maps that shows changes in some spatial phenomenon over time

toggling: a feature on an interactive map that enables users to hide or reveal map features

topography: description of the shape of the earth's surface; a topographic map is a reference map that shows the land and other features of the earth

topology: a branch of mathematics concerning the study of geometric properties and spatial relations among spatial features

tracking: the distance between individual letters in text

transparency: a graphical effect that makes objects transparent, allowing some of the color from graphical objects underneath to show

trichromatic theory: that the idea that we perceive color using three types of optical cells that detect different levels of radiation at different frequency ranges; the brain uses the combination of detected intensities of radiation from the frequency ranges to build the perception of color

type of symbol: a point, line, polygon, or volume used to represent geographic features on a map

typeface: synonymous with font; the style and shape of the letters

typestyle: variations of fonts, such as bold and italics

typography: the study of text and its appearance

Universal Polar Stereographic (UPS): An accurate international coordinate system developed by the US military to map areas near to the poles, based on azimuthal stereographic projections; this coordinate system complements the Universal Transverse Mercator (UTM) coordinate system

Universal Transverse Mercator (UTM): an accurate international coordinate system developed by the US military, based on zone-based iterations of a transverse Mercator projection

user interface design: designed focused around the ease of use and enjoyment of users of a computer application

variable resolution: a computer device on which the screen resolution can be altered

variable scale: a type of graphical scale that shows how the scale changes across different portions of the map

vector spatial data: spatial data built from lists of coordinate pairs

visual contrast: the contrast between map features, elements, and symbols with their backgrounds, other elements, and the layout in general

visual direction: the "center of gravity" in the layout of a map; well-balanced maps have a center of gravity near the center of the layout

visual hierarchy: the visually communicated structure of the hierarchy of importance in a map; elements high in the visual hierarchy are the first to draw the reader's attention

visual variables: types of systematic variations in map symbology, such as size or color, to communicate information

visual weight: the extent to which a map element dominates the view on a map; thick lines have a higher visual weight than thinner ones

visualization: the general idea of "making data visible" by translating abstract concepts to a human scale in something that can be viewed (either physically in a diagram or mentally)

volume symbols: cartographic symbols that appear as volumes on a map; volume symbols can be expressed as true volume in three-dimensional maps or mimicked in two-dimensional maps; they can represent topography or a variety of other aerial physical or thematic phenomena

volunteered geographic information (VGI): crowdsourced spatial data and the systems built around facilitating its use and dissemination

web mapping: mapping that occurs primary over Internet media, also called "web mapping"

white space: synonymous with "negative space"

widgets: components of a digital map that enable users to interact with the application

World Geodetic System of 1984 (WGS84): a general reference ellipsoid for mapping the entire surface of the earth, which is heavily used in geographic positioning system (GPS) and geospatial intelligence mapping

xls (file format): *Microsoft Excel* files, designed for use in Microsoft's primary spreadsheet application; Excel files can often be read by applications designed to work with tabular data

Index